Experimental Brain Research Series 23

R. J. do Carmo (Ed.)

Spreading Depression

With 50 Figures

Springer-Verlag
Berlin Heidelberg New York
London Paris Tokyo
Hong Kong Barcelona
Budapest

Prof. Romualdo José do Carmo
Universidade Federal do Rio de Janeiro
Centro de Ciências da Saude
Instituto de Biofísica Carlos Chagas Filho
CCS-Bloco G
Cidade Universitaria
21949 Rio de Janeiro RJ
Brazil

Cover illustration:
Pictures of SD optical changes in chick eye cup, taken at different time intervals. (Photo by C. J. Struchiner)

ISBN-13:978-3-642-77553-6 e-ISBN-13:978-3-642-77551-2
DOI: 10.1007/978-3-642-77551-2

Library of Congress Cataloging-in-Publication Data
Spreading depression / R. J. do Carmo (ed.). p. cm. -- (Experimental brain research series ; 23)
Includes bibliographical references and index.
ISBN-13:978-3-642-77553-6
1. Spreading cordical depression. I. Series.
[DNLM: 1. Electrophysiology--congresses. 2. Spreading Cortical Depression--congresses. W1 EX4855 v.23 / WL 397 S768]
RC347.S67 1992 616.8--dc20 DNLM/DLC for Library of Congress 92-2882 CIP

Softcover reprint of the hardcover 1st edition 1992

25/3145 - 5 4 3 2 1 0 - Printed on acid-free paper

Preface

During the past five decades, several groups of scientists throughout the world have endeavoured to advance the frontiers of knowledge on spreading depression (SD), the "phenomenon of Leão," as it was called after its discoverer, the Brazilian physiologist Aristides Leão.

Spreading depression was first described in studies carried out in the domain of so-called experimental epilepsy. The relationship of the phenomenon to seizure marches has been noticed since the early studies and its possible significance to clinical conditions such as migraine, brain concussion, and ischemia was also soon recognized. In the past decade, this line of investigation has gained new impetus.

This volume is a collection of ten original articles written by authorities in the field, who have brought together historical issues, new ideas about fundamental mechanisms of the reaction, its relationship to clinical entities and conditions of protein deficiency, and the new experimental approach of recording magnetic signals during SD in vivo as well as in vitro.

It is very appropriate that this volume should be dedicated to Prof. Hiss Martins-Ferreira on the occasion of his retirement, which, however, is not a de facto retirement, since he will continue to devote his fruitful scientific life to the study of SD.

It is most gratifying to have the benefit of the active participation of Prof. Aristides Leão, who, as the discoverer of SD, is ultimately responsible for this meeting. The moment is also appropriate to remember the names of Wade Marshall and Anthonie Van Harreveld, who contributed so much to the knowledge of the phenomenon.

We are grateful to Fundação de Amparo à Pesquisa do Estado do Rio de Janeiro (FAPERJ), Fundação José Bonifácio da Universidade Federal do Rio de Janeiro

(FUJB/UFRJ), Conselho Nacional de Desenvolvimento Científico e Tecnológico (CNPq), IBM of Brazil, Programa Avançado de Neurociências da Universidade Federal do Rio de Janeiro (PAN/UFRJ), and to the Health Foundation, whose financial support made it possible for this conference to take place; to Mrs. L. E. Serrano do Carmo, for her assistance; and to Prof. M. Sorensen for revising some of the manuscripts.

While this volume was in press, Professor Otto Creutzfeldt passed away – a most significant loss for the neurosciences troughout the world.

R. J. do Carmo

Contents

List of Contributors

ANTHONY R. GARDNER-MEDWIN
Department of Physiology, University College London,
London WC1E 6BT, U.K.

ARISTIDES A. P. LEÃO
Departamento de Neurobiologia, Instituto de Biofísica Carlos Chagas Filho,
Univrsidade Federal do Rio de Janeiro,
Ilha do Fundão, 21941, RJ, Brazil

CELSO REZENDE FERREIRA-FILHO
Departamento de Fisiologia, Instituto Biomédico da Universidade
Federal Fluminense, Rua Professor Ernani Mello,
101, 24210, Niterói, RJ, Brazil

CHARLES NICHOLSON
Department of Physiology and Biophysics,
New York University Medical Center,
550 First Ave., New York, NY 10016, USA

GEORGE G. SOMJEN
Divison of Physiology, Duke University Medical Center,
Durham, North Carolina 27710, USA

HISS MARTINS-FERREIRA
Departamento de Neurobiologia, Instituto de Biofísica Carlos Chagas Filho,
Universidade Federal do Rio de Janeiro, Ilha do Fundão, 21941, RJ Brazil

JAN BUREŠ
Institute of Physiology, Czechoslovak Academy of Sciences,
Videnska 1083, 14220 Prague 4 - Krr Czechoslovakia

J. E. Cabral Filho
Departamento de Nutrição, Universidade Federal de Pernambuco, 50739, Cidade Universitaria, Recife, PE, Brazil

J. Jing
Divison of Physiology, Duke University Medical Center, Durham, North Carolina 27710, USA

Ludmila V. Vinogradova
Institute of Higher Nervous Activity and Neurophysiology, Academy of Sciences, 5a Butlerov Str., Moscow 117865, Russia

Martin Lauritzen
Department of General Physiology, University of Copenhagen and Laboratory of Clinical Neurophysiology, Rigshospitalet, Copenhagen, Denmark

Naide R. Teodosio
Departamento de Nutrição, Universidade Federal de Pernambuco, 50739, Cidade Universitária, Recife, PE, Brazil

N. A. Gorelova
Institute of Higher Nervous Activity and Neurophysiology, Academy of Sciences, 5a Butlerov Str., Moscow 117865, Russia

Oscar Herreras
Divison of Physiology, Duke University Medical Center, Durham, North Carolina 27710, USA

Romualdo José do Carmo
Departamento de Neurobiologia, Instituto de Biofísica Carlos Chagas Filho, Universidade Federal do Rio de Janeiro, Ilha do Fundão, 21941, RJ, Brazil

Rubem Carlos Araujo Guedes
Departamento de Nutrição, Universidade Federal de Pernambuco, 50739, Cidade Universitária, Recife, PE, Brazil

Valeria I. Koroleva
Institute of Higher Nervous Activity and Neurophysiology, Academy of Sciences, 5a Butlerov Str., Moscow 117865, Russia

Yoshio C. Okada
Departments of Neurology and Physiology, University of New Mexico School of Medicine and V A Medical Center, Albuquerque, NM 87108, USA

Spreading Depression: Some Reminiscences

A. A. P. Leão

Departamento de Neurobiologia, Instituto de Biofísica Carlos Chagas Filho, Universidade Federal do Rio de Janeiro, Ilha do Fundão, 21941, RJ, Brasil

Let me begin by reminding you that the description, in extenso, of spreading depression appeared in the November 1944 issue of the *Journal of Neurophysiology* (Leão 1944), and that additional data, obtained by Robert Morison and myself, were published in the next issue, i.e., January 1945, of the same journal (Leão and Morison 1945).

At that time we were somewhat anxious to find out whether anyone else had become interested in the phenomenon. We had to wait until 1947. In that year, when I was about to submit my paper on the slow potential change of spreading depression for publication (Leão 1947), I received a letter from Benjamin Libet, of the University of Chicago, which started thus:

Dr. Kenneth Cole and Dr. Ferreira have told Dr. Gerard and me about your findings of DC changes in conjunction with setting up the spreading depression waves which you have described. We were delighted to hear this for we had also been studying such changes and apparently our results are in agreement with yours.

You are all acquainted with the name Kenneth Cole. This "Dr. Gerard" is the, also well known, neurophysiologist Ralph Gerard, who was then professor of physiology at the University of Chicago. And this "Dr. Ferreira" is our Hiss Martins-Ferreira, who was then on a fellowship in Kenneth Cole's laboratory. Libet's letter went on to describe the potential changes that they had observed during spreading depression. The description was, in fact, essentially in agreement with mine, and this was stated in a note appended to my paper. In a subsequent letter Dr. Libet said that he was engaging in a large project on nerve metabolism, and was not, at that time, going to proceed with the work on the DC changes of spreading depression, although he continued to be interested in them. In fact, only much later, in 1962, did he publish, with Gerard, a paper on this subject (Libet and Gerard 1962). Nothing else on spreading depression turned up in 1947. In 1948, a brief but noteworthy paper appeared in the *Transactions of the American Neurological Association*. The author, Dr. Francis Echlin, of New York, reported some experiments on spreading depression that he had performed on rabbits, and after considering what was known of concussion – in particular the observations of Williams and Denny-Brown on experimental concussion – he concluded with this stimulating sentence: "It is suggested that the EEG findings in concussion may be largely explained on the basis of a spreading depression of electrical activity initiated from innumerable points in the cerebral cortex simultaneously" (Echlin 1948). Echlin was the only newcomer in 1948, but, as far as I know, he did not pursue the subject further.

In 1949 appeared the first publication by Wade Marshall, whose studies, which continued for many years, were to bring about a fundamental advance in the knowledge of spreading depression. This 1949 publication was the abstract of a communication to the *American Physiological Society* by Marshall and his collaborators Essig and Dubroff, at the United States' National Institutes of Health. It reported the effects of changing the temperature of a petrolatum pool over the exposed cortex of rabbits and monkeys (Marshall et al. 1949).

In the next decade, the 1950s, much happened, but I shall not go into details. What was most important was the entrance into the field of Anthonie Van Harreveld, in the United States, and Jan Bureš, in Europe, who were to become leaders in the further investigation of spreading depression. And also the entrance of Hiss Martins-Ferreira, of which I take pride in having been the inducer. His first publication in the field was our joint paper on the electrical impedance changes accompanying spreading depression, published in 1953 (Leão and Martins-Ferreira 1953).

In the late 1950s and early 1960s it was established that spreading depression may occur in several neuronal aggregations of the brain of warm- and cold-blooded vertebrates. Among these neuronal aggregations was the excised amphibian retina, as demonstrated by Gouras (1958). This was a case apart, not only because the nervous portion of the retina is part of the brain only in early embryonic stages, but also because the spreading depression was observed in vitro, and was accompanied by a change in the optical properties of the tissue, which made the entire extent of the spread of the phenomenon easily visible.

Now, Martins-Ferreira has always had a predilection for analytical methods, and this innate bent was reinforced in his stays with Kenneth Cole, at the University of Chicago, as already mentioned, and with Torsten Teorell, at the University of Uppsala – two men of outstanding analytical skill. It was therefore to be expected that the retina in vitro would have a special attraction for him. There was the possibility of maintaining the preparation in the same state for at least a few hours, which would reduce the variability, or perhaps better, as Wade Marshall used to say, the capriciousness, of the spreading depression. Thus, he at once started experimenting.

The original preparation was simple. The eyeball was sectioned along the equator, the vitreous humor in the posterior segment was drawn off, and this segment, comprising most of the retina, choroid and sclera, was placed in a bath of Ringer's solution. It was therefore like a hemispheroidal cup, the freely accessible inner surface of which being the anterior surface of the retina.

Martins-Ferreira soon found out that, for his purpose, the avian eye provided better material than the amphibian eye. The vitreous humor is less viscous, and so more easily drawn off from the posterior segment of the sectioned eyeball. Besides, the avian retina is entirely devoid of blood vessels. In the live bird, therefore, it is only through the retinal surfaces that exchange of substances takes place, and in the isolated preparation the bathing solution may be brought into direct contact with the anterior surface of most of the retina. The eye of the chick of the domestic fowl was found to be satisfactory, and a Ringer's solution of a particular composition, to which glucose was added, proved to be most suitable for the study of the spreading depression.

As regards invariance, Martins-Ferreira had got what he wanted: a preparation in which the spreading depression presented itself for hours with a regularity and repeatability in no way inferior to that of the responses of peripheral nerve or muscle.

Still at a preliminary stage, it was ascertained that spreading depression could also be provoked in fragments of retina detached from the eyecup preparation. If the pigment layer was stripped from a fragment, a preparation consisting of only the nervous portion of the retina was obtained.

A definite need was an accurate record of the optical changes accompanying retinal spreading depression, and this was the first problem that Martins-Ferreira addressed. In this endeavor he was fortunate in having as collaborator Gustavo de Oliveira Castro, an expert in microscopy, endowed with exceptional ingenuity. The changes are in the intensity of the light scattered within the nervous portion of the retina, and, since they move along this lamina of tissue, an obvious requisite is that any measurement be confined to a small part of it. In a short time, and with pieces of equipment picked up here and there in the laboratory, they built up a microphotometric system capable of measuring the intensity of the light coming out exclusively from a circular area of the retinal surface only 45 μm in diameter (Martins-Ferreira and Oliveira Castro 1966).

At the incitation of Bureš, during a brief visit in 1971, a new preparation was worked out, with the purpose of producing the so-called circling spreading depression. A circular incision was made in the eyecup preparation, separating a posterior callote and an anterior ring of retinal tissue. If a spreading depression was locally provoked in this ring, it spread in opposite directions, and was annihilated when the two wave fronts met. However, it was found that if the spread was blocked, by anodic polarization, in one of the wave fronts, and the tissue in this region had recovered by the time the other wave front reached it, the spreading depression passed through it and continued going round and round the ring for hours on end (Martins-Ferreira et al. 1974).

Advancing cautiously, onward step by step, Martins-Ferreira is performing a most comprehensive study of spreading depression in the excised retina. Of the topics so far investigated, mention may be made of the concomitant mechanical changes; changes in the field potential, activity of extracellular ions, and volume of the extracellular space, at different layers of the retina; the effect of changes on the concentration of potassium and in the temperature of the bathing solution; the role of glucose; the participation of cholinergic and GABAergic systems; and the effects of convulsant and anticonvulsant drugs. Recent results of experiments with unsaturated fatty acids will be presented at this symposium.

I shall single out a piece of work, which shows how varied the output of his laboratory has been. Impressed, like many of us, with the similarity, in various features, between spreading depression and the self-propagating Belousov-Zhabotinskii reaction, he thought that some experiments would be of interest in this respect. In fact, under his supervision a student, who was well grounded in physical chemistry, succeeded in producing a circling Belousov-Zhabotinskii reaction in a cellulose acetate ring of nearly the same perimeter as the aforementioned ring of retinal tissue. By varying the concentration of one of the reagents in the solution absorbed in the ring, a system was obtained in which the reaction did not occur spontaneously, but could be locally initiated by touching the ring with a hot wire. The reaction started to move in opposite directions, and was annihilated when the two fronts met.

It could, however, be stopped in one of the fronts by applying a blade of iron across its path for a few seconds. A little later the other front was able to pass through this region, and this done, the reaction continued going round and round the ring (França Faria 1980).

Martins-Ferreira has supervised the work of many graduate students. A recently concluded work – an MSc thesis – demonstrated that spreading depression may also be provoked in the retina in situ. The experiments were carried out on anesthetized hens, and the optical changes in the fundus of the eye were viewed through the pupil (Streit 1990).

Except for the work of instructing and training new students, the research work of Martins-Ferreira, since 1977, has been carried out in collaboration with Romualdo José do Carmo, a happy association, from which more advances can be expected, for I am sure that in retirement Martins-Ferreira will continue to be as active in research as ever.

References

Echlin, F (1948) Spreading depression of electrical activity in the cerebral cortex following local trauma and its possible role in concussion. Trans Am Neuro Assoc, 73: 199-202

França Faria MA (1980) Depressão alastrante e reação de Belousov-Zhabotinsky. MSc Thesis, Institute Biophysics, Federal University of Rio de Janeiro

Gouras P (1958) Spreading depression of activity in amphibian retina. Am J Physiol 195: 28-32

Leão A A P (1944) Spreading depression of activity in the cerebral cortex. J Neurophysiol 7: 359-390

Leão A A P, Morison R S (1945) Propagation of spreading cortical depression. J Neurophysiol 8: 33-45

Leão A A P (1947) Further observations on the spreading depression of activity in the cerebral cortex. J Neurophysiol 10: 409-414

Leão A A P, Martins-Ferreira H (1953) Alteração da impedância elétrica no decurso da depressão alastrante da atividade do córtex cerebral. An Acad Bras Ciênc 25: 259-266

Libet B, Gerard R W (1962) An analysis of some correlates of steady potentials in mammalian cerebral cortex. Electroencephalogr Clin Neurophysiol 14: 445-452

Marshall W H, Essig C F, Dubroff S J (1949) Essential factors involved in spreading cortical depression. Am J Physiol 159: 579

Martins-Ferreira H, Oliveira Castro G (1966) Light-scattering changes accompanying spreading depression in isolated retina. J Neurophysiol 29: 715-726

Martins-Ferreira H, Oliveira Castro G, Struchiner C J, Rodrigues P S (1974) Circling spreading depression in isolated chick retina. J Neurophysiol 37: 773-784

Streit D S (1990) Depressão alastrante retiniana "in vivo". MSc Thesis, Department of Medicine (Ophtalmology), Federal University of Rio de Janeiro

Cortical Spreading Depression as a Migraine Mechanism: Clinical and Experimental Aspects

M. Lauritzen

Department of General Physiology and Biophysics, University of Copenhagen, Laboratory of Clinical Neurophysiology, Rigshospitalet, Copenhagen, Denmark

The Theories

Migraine is a common disorder affecting about 10% of the adult population. Paroxysmal headaches, typically associated with focal neurological symptoms and signs (classic migraine), are the key symptoms. Classic migraine attacks have a characteristic course: at first the patients experience transient focal neurological symptoms with a characteristic pattern of evolution (see below), spontaneously remitting within 30-60 min. At the same time as or after the focal symptoms have disappeared, the headache ensues, which most often is unilateral and throbbing.

To explain this characteristic sequence of events there are at present two theories: one is the "vascular" theory, which assigns the primary event in migraine to a disturbance of cerebral vascular function. The focal symptoms are supposed to be causally related to transient constriction of a cerebral artery, and the headache to a sterile inflammatory reaction around the walls of dilated cephalic vessels (Wolff 1963).

Several arguments are in favor of such a concept. Firstly, the characteristic of throbbing pain in migraine suggests vasodilatation. Secondly, headache is a common concomitant of other diseases involving blood vessels such as stroke, subarochnoid hemorrhage, arterial hypertension, and temporal arteritis. Thirdly, the brain itself is insensitive to pain, while pain-sensitive fibers have been identified at its ventral surface corresponding to the larger vessels and associated structures. Fourthly, the peripheral vasoconstrictor ergotamine is a potent drug in the alleviation of the headaches (Wolff 1963). The "vascular" theory has dominated the migraine field for decades.

The second theory assigns the primary event in migraine to a disturbance of neuronal function (Gowers 1907). Firstly, the regular recurrence of an unprovoked, transient disturbance of brain function is reminiscent of epilepsy more than of any known disorder of the peripheral vasculature. Secondly, the focal symptoms in migraine develop in a characteristic "creeping" fashion. The slow march of symptoms is commonly used to differentiate migraine and epilepsy. For example, many patients with classic migraine have a characteristic disturbance of vision, a scintillation-scotoma which develops symmetrically in the visual fields, starting at the center of the field of vision and propagating to the peripheral (temporal) parts within about 10-15 min. Visual function returns to normal within another 10-15 min. The symmetrical development suggests a cortical localization of the underlying disturbance. The slow progression indicates a wave of excitation in the primary visual cortex moving at the speed of 3 mm/min, followed by a longer period of inhibition (Fig. 1) (Lashley 1941). Similar considerations and calculations can be made when somatosensory or somatomotor instead of visual symptoms develop. The very orderly development of symptoms makes a vascular origin a remote possibility (Gowers 1907), while a primary disturbance of cortical nerve cell function, probably cortical spreading depression (CSD), is a more attractive explanation (Milner 1958).

The two theories may be unified if on one hand a CSD develops during the onset of the migraine attack, and if on the other hand CSD triggers changes of the perivascular microenvironment which activate pain-sensitive fibers, in turn causing pain (Lauritzen 1987a, b). This paper presents data which suggest such a theory.

Brain Blood Flow in Humans

Regional cerebral blood flow (rCBF) studies during acute migraine have revealed (reviewed in Lauritzen 1987a):

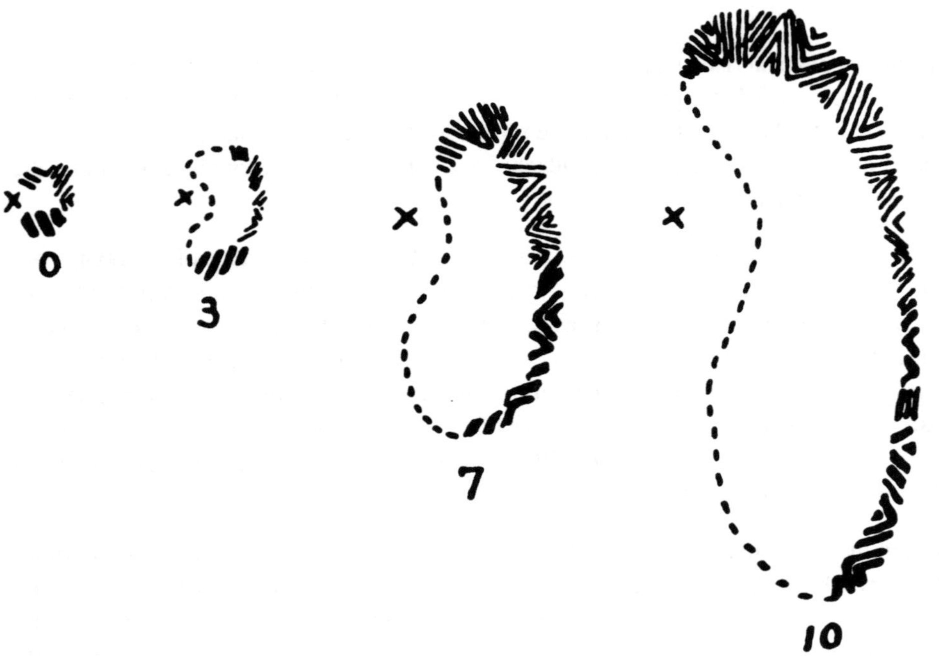

Fig. 1. Successive maps of a scintillation scotoma showing characteristic distribution of the fortification figures. In each case *the asterisk* indicates the fixation point. Knowledge of the retinotopic organization of the visual cortex allowed Lashley (1941) to calculate the speed of propagation of the excitation-depression wave to be approximately 3 mm/min. (From Lashley 1941)

1. Regional cerebral blood flow decreases during the initial part of the attack in the posterior part of the brain. Subsequently, the region of hypoperfusion expands in an irregular fashion and to a varying extent in individual patients at a rate of 2-3 mm/min.

2. Spread of the hypoperfusion does not respect territories of supply of the large arteries, but follows the cortical surface. Major cytoarchitectonic boundaries, e.g., the central sulcus, and larger foldings, e.g., the lateral sulcus, remain uncrossed. The hypoperfusion propagates from temporal and parietal cortices to frontal regions along the insular region and the parietal operculum, respectively.

3. The rCBF changes are localized to the cortex.

4. Blood pressure autoregulation is preserved, i.e., rCBF is maintained constant despite changes in arterial blood pressure. This indicates cortical arterioles as the site of increased resistance with residual capacity to constrict and dilate in response to changes of perfusion pressure. Vasospasm of a large artery would cause compensatory dilatation of the arterioles if severe enough to reduce rCBF. Hence, increased blood pressure would lead to rCBF increases in a pressure-passive arteriolar bed.

Since the systemic factors (blood gases, hematocrit) remained constant, local factors were suspected of causing the increased vascular tone.

5. Vascular reactivity to mental tasks and CO_2 reactivity are impaired in hypoperfused parts, but present in neighboring noninvaded tissue, suggesting decreased responsiveness to perivascular chemical stimuli.

6. Focal symptoms appear at some time point during the (early) spread of the hypoperfusion, but cease while the hypoperfusion continues to spread. From closely spaced studies it has become clear that the reduced rCBF persists during the headache period. Eventually, after about 5-6 h and lasting for up to 24 h, a hyperperfusion is observed in the brain regions which were previously hypoperfused (Lauritzen 1987a; Andersen et al. 1987).

7. The rCBF reduction amounts to about 20% – 25%. Recently, it has been claimed that the reduction of rCBF is larger than originally believed due to errors of measurement inherent in our technique (Skyhøj Olsen et al. 1987). In brief, the reduced rCBF in the hypoperfused regions may be overestimated due to Comptom scattered radiation from the neighboring normoperfused regions (Skyhøj Olsen et al. 1987). However, it is still unclear whether, if at all, this source of error plays any significant role in our conclusions (Kronborg et al. 1990), and the subject will not be discussed further for the present.

In conclusion, the time base of the focal symptoms and the reduced rCBF are clearly different. This is a strong argument against the "vascular" theory of migraine. The vascular changes are consistent with a slowly propagating, extinguishable cortical process: spreading at a rate of 2-3 mm/min, constricting cortical arterioles, and causing focal neurological symptoms while expanding. The reduced rCBF appears to be an epiphenomenon, but may under certain (unknown) conditions contribute to the persistence of neurological deficits.

The headache in turn is not associated with increased rCBF, and there is no evidence of dilated cerebral arteries during the attack. It is possible that the spreading cortical disturbance, which has sufficient impact to constrict the arterioles, also activates the pain-sensitive nerve fibers around the larger arteries, but this remains to be shown.

There is only one known cortical disturbance that fits into the above description of rCBF changes during migraine attacks: the cortical spreading depression of Leão (Leão 1944).

Brain Blood Flow in Rats

Regional cerebral blood flow changes during CSD occur in a regular fashion (reviewed in Lauritzen 1987a). There is probably no change in blood flow before the nerve cells depolarize or during the onset of depolarization. During return to normal of the ionic changes, a flow rise is sometimes observed, especially if the basic level of flow is depressed due to the use of anesthetics or other blood flow lowering drugs. When present, the flow rise lasts about 1-2 min, being rapidly followed by a 20%–30% flow reduction. The rCBF reduction is strictly localized to the cortex while non-invaded tissues have a normal flow, i.e., the arteriolar vasoconstriction is due to the influence of local factors. While vasoconstriction is clearly observed in the rat, the cat displays a continued arteriolar vasodilation. Thus, there is species differences with respect to the vascular reaction (Wahl et al. 1987).

The hypoperfusion in rats persists for about 2 h. Blood pressure autoregulation is preserved in the entire brain while the responsiveness to changes of $PaCO_2$ and locally applied chemical stimuli is markedly impaired in hypoperfused regions. The cerebral metabolic rate of glucose remains normal despite the low flow. Thus, the rCBF reduction is not severe enough to impede cortical glucose consumption (Lauritzen and Diemer 1986).

In summary, CSD in rats is associated with a brief initial flow rise under certain conditions. Following CSD, rCBF declines and remains reduced for about 2 h. The reduced rCBF is caused by arteriolar vasoconstriction and associated with preserved autoregulation and diminished chemical responsiveness. This pattern of rCBF changes replicates almost point by point the rCBF changes associated with migraine.

Energy Metabolism

Our knowledge of the changes of energy-rich phosphate compounds, and oxygen and glucose metabolism, during migraine attacks is minimal. This is due to the inherent difficulties in investigating short-lasting paroxysmal disorders, where the disturbance of metabolic function may have disappeared at the time the patient is referred to, or comes to, the clinic. Furthermore, the use of the positron emission tomography (PET) scanner requires preparatory procedures, and the relevant radiochemical compound may not be available when the patient is at the clinic. Nevertheless, one PET study has been carried out, and the results showed a normal cerebral metabolic rate of oxygen in the hypoperfused regions of a patient with classic migraine at the time of neurological deficits (Herold et al. 1985). This result is incompatible with cerebral ischemia. Therefore, the PET data support the idea that cerebral ischemia is not causally linked to the existence of focal deficits in migraine.

Magnetic resonance imaging (MRI) techniques have been applied to migraine patients at 3-48 h after the start of the attack using a surface coil to obtain signals from a 4-cm spherical source of the cerebrum. The preliminary results showed a decrease in the phosphocreatine (PCr) concentration, an increase in inorganic phosphate, while ATP and intracellular pH remained constant. The absence of pH changes was taken as an argument against the "vascular" theory since pH was found to be reduced in the affected region in almost all patients with a stroke or transient ischemic attack (TIA) (Welch et al. 1989).

The intracellular pH during and following CSD is unknown. We do know, however, that the concentration of energy metabolites changes markedly at the wave front, lasting for a few minutes: PCr decreases by about 38% while ATP remains constant (Lauritzen et al. 1990). Thus, the changes in PCr and ATP during CSD and classic migraine attacks are similar, suggesting an increased turnover rate of these compounds, but no evidence of ischemia!

Magnetoencephalogram

The magnetoencephalogram (MEG) signal associated with CSD was obtained for the first time under in vitro conditions using the isolated turtle cerebellum and a one-channel DC-coupled superconducing quantum interference device (SQUID) (Okada et al. 1988). The results revealed a well-defined signal in the picotesla range, raising the possibility of noninvasively detecting CSD in the human cerebral cortex in association with migraine attacks. Thus, the field appeared to be strong enough to be measured outside the head in single trials, provided that the strength of the magnetic field generated by the human neocortex was within the same order of magnitude as the field recorded from the turtle cerebellum.

Promising results have been obtained with a seven-channel AC-coupled MEG system in a group of migraine patients (Tepley et al. 1990). The MEG signals obtained share important features with the signals obtained with the same equipment from rabbits during CSD (Gardner-Medwin et al. 1990). However, we are still ignorant of the specificity of the observed signals. Therefore, further confirmation of the results is needed. Undoubtedly the MEG technique has the potential to clarify the controversies regarding the "vascular" versus the CSD theory. But there is also no doubt that a very specific signal which spreads is required if we are to be convinced that a wave of CSD has been recorded.

Behavior and Symptoms

Single unilateral waves of CSD induce remarkably modest signs of cortical dysfunction in rats. Contralateral sensory neglect and motor impairment of the contralateral forepaw are the most reliable neurological signs lasting for about 15-30 min, i.e., much shorter than the blood flow reduction. It is unknown whether CSD is an aversive or a pleasant stimulus. It is also unknown how CSD might cause headache in rats, nor can it be judged whether the rat has headache or not (Bureš et al. 1974, 1984).

Behavioral changes due to remote effects of CSD on subcortical centers are usually longer lasting than the cortical deficits. Signs of possible hypothalamic origin include abnormal thermoregulation and feeding behavior, water retention, and ovarian hormone secretion.

In conclusion, CSD in rats is accompanied by transient contralateral somatosensory and motor deficits, and signs of involvement of the hypothalamus. Attacks of migraine with neurological deficits have similar symptoms with a similar time course. The analysis can hardly be performed more rigorously since a major discrepancy between the animal experiments and the human condition is the involvement of the whole hemicortex in the rat, while during the classic migraine attack most often only regions located in the posterior parts of the brain are involved (Lauritzen 1987a, b).

Perspectives

In conclusion it appears likely that CSD is involved in the sequence of events that leads to acute migraine episodes. This changes our view of migraine from that of episodic changes of blood vessel function to a well-defined disturbance of cortical neuronal function, cortical spreading depression (Fig. 2).

Though our understanding of CSD is still incomplete, the recent development and application of drugs directed against glutamate receptors have given us important clues to events at the synaptic level which are presumed to be important in the production of CSD.

Van Harreveld (1959) was the first to propose that glutamate had an active role in CSD. A few years later it was shown that N-methyl-D-aspartate (NMDA) was 100 times more potent than glutamate in triggering CSD (Curtis and Watkins 1961). In another series of experiments it was found that glutamate is released during CSD, and that various glutamate antagonists block CSD in the chicken retina (Van Harreveld and Fifková 1970, 1971; Van Harreveld 1984). Recently it has been shown that:

1. Hippocampal and cerebellar CSD occur spontaneously in Mg-free medium, a condition which is supposed to facilitate the entrance of calcium ions through the NMDA-operated membrane channel, suggesting a role for this receptor-channel complex in CSD (Mody et al. 1987; Lauritzen et al. 1988).

2. Cortical spreading depression is blocked by D-α-aminoadipate, ketamine, MK-801, amino-5-phosphonovalerate (APV), and amino-7-phosphonoheptanoate (APH), i.e., competitive and noncompetitive NMDA antagonists (Van Harreveld 1984; Goroleva et al. 1987; Marannes et al. 1988; Lauritzen et al. 1988).

3. Antagonists of non-NMDA receptors are much less effective as blockers of CSD (Lauritzen and Hansen 1991).

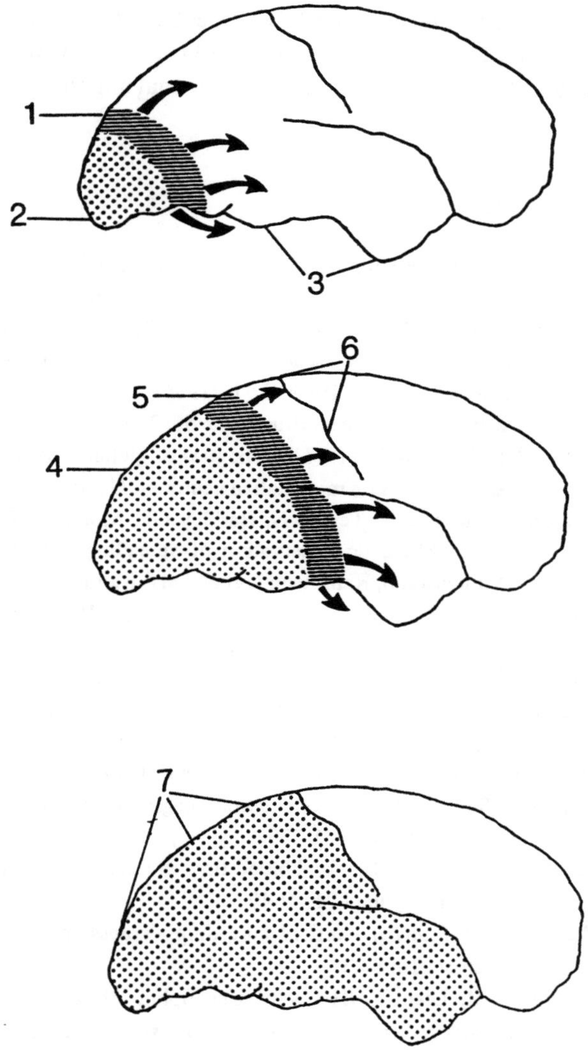

Fig. 2. Hypothesis of development of a classic migraine attack based on aspects of CSD and migraine summarized in the text. The figures represent lateral views of the human brain at different time intervals after the start of the attack, spaced by approximately 30 min. *Dotted area* represents the region of reduced rCBF; *striped area* represents the region of neuronal depolarization during the first minutes of CSD; *arrows* indicate the direction of progression of CSD. *1*, initially during a classic migraine attack a CSD is elicited at the occipital pole, spreading anteriorly at the lateral, mesial, and ventral sides of the brain. At the CSD wave front, transient ionic and metabolic disequilibria trigger perturbed neuronal function, rCBF changes, and neurological symtoms. *2*, following CSD, cortical rCBF decreases by 20%–30% for 2-6 h. *3*, rCBF in regions not invaded by CSD remains normal until encountered by CSD. *4*, the region of reduced rCBF expands as the CSD moves anteriorly. *5*, somatosensory symptoms from the extremities appear when the CSD invades the primary sensory cortex at the postcentral gyrus. *6*, CSD usually stops on reaching the central sulcus, but in many patients it does not even spread this far. The ventral spread of CSD causes activation of pain-sensitive fibers and headache. *7*, full-scale attack; the CSD has stopped and is now detectable as a persistent reduction of cortical rCBF. At this time the patient suffers from headache, but has no focal deficits. (From Lauritzen 1987b)

The results indicate that NMDA receptor activity is of crucial importance for the production of CSD. If indeed CSD proves to be the causative mechanism of migraine, NMDA antagonists could be the next type of rational migraine therapy.

References

Andersen AR, Friberg L, Skyhøj Olsen T, Olesen J (1987) Delayed hyperemia following hypoperfusion in classic migraine. Arch Neurol 44: 156-161

Bureš J, Burešová O, Křivánek J (1974) The mechanism and applications of Leão's spreading depression of electroencephalographic activity. Academic, New York

Bureš J, Burešová O, Křivánek J (1984) The meaning and significance of Leão's spreading depression. An Acad Bras Ciênc 56: 385-400

Curtis DR, Watkins JC (1961) Analogues of glutamic acid and gamma-amino-*n*-butyric acids having potent actions on mammalian neurones. Nature 191: 1010-1011

Gardner-Medwin AR, Tepley N, Barkley GL, Moran J, Nagel-Leiby S, Simkins RT, Welch KMA (1990) Magnetic observation of spreading cortical depression in anesthetized rabbits. In: Williamson S, Kaufmann L et al. (eds) Biomagnetism. Plenum, New York, pp 323-326

Goroleva NA, Koroleva VI, Amemori T, Pavlik V, Bureš J (1987) Ketamine blockade of cortical spreading depression in rats. Electroencephalorg Clin Neurophysiol 66: 440-447

Gowers WR (1907) The borderland of epilepsy. Churchill, London

Herold S, Gibbs JM, Jones AKP, Brooks DJ, Frachiowak RSJ, Legg NJ (1985) Oxygen metabolism in migraine. J Cereb Blood Flow Metab 5 [Suppl 1]: S445-446

Kronborg D, Dalgaard P, Lauritzen M (1990) Letter to the editor. Arch Neurol 47: 124-125

Lashley KS (1941) Patterns of cerebral integration indicated by the scotomas of migraine. Arch Neurol Pschychiatry 46: 331-339

Lauritzen M (1987a) Cerebral blood flow in migraine and cortical spreading depression. Acta Neurol Scand 113 [Suppl 76]: 9-40

Lauritzen M (1987b) Cortical spreading depression as a putative migraine mechanism. Trends Neurosci 10: 8-13

Lauritzen M, Diemer NH (1986) Uncoupling of cerebral blood flow and metabolism after single episodes of cortical spreading depression in the rat brain. Brain Res 370: 405-408

Lauritzen M, Rice M, Okada Y, Nicholson C (1988) Spreading depression by activation of glutamate receptors in turtle cerebellum *in vitro*. Brain Res 475: 317-327

Lauritzen M, Hansen AJ, Kronborg D, Wieloch T (1990) Cortical spreading depression is associated with arachidonic acid accumulation and preservation of energy charge. J Cereb Blood Flow Metab 10: 115-122

Lauritzen M, Hansen AJ (1991) The effect of glutamate receptor blockade on anoxic depolarization and cortical spreading depression. J Cereb Blood Flow Metab (in press)

Leão AAP (1944) Spreading depression of activity in the cerebral cortex. J Neurophysiol 7: 359-390

Marannes R, Willems R, De Prins E, Wauquier A (1988) Evidence for a role of the N-Methyl-D-aspartate receptor in cortical spreading depression in the rat. Brain Res 457: 226-240

Milner B (1958) Note on a possible correspondence between the migraine scotomata and spreading depression. Electroencephalogr Clin Neurophysiol 10: 705

Mody I, Lambert JDC, Heinemann U (1987) Low extracellular magnesium induces epileptiform activity and spreading depression in rat hippocampal slices. J Neurophysiol 57: 869-888

Skyhøj Olsen T, Friberg L, Lassen NA (1987) Ischemia may be the primary cause of neurological deficits in classic migraine. Arch Neurol 44: 156-161

Okada YC, Lauritzen M, Nicholson C (1988) Magnetic field associated with spreading depression: a model for the detection of migraine. Brain Res 442: 185-190

Tepley N, Barkley GL, Moran JE, Simkins RT, Welch KMA (1990) Observation of cortical spreading depression in migraine patients. In: Williamson S, Kaufmann L et al.(eds) Biomagnetism. Plenum, New York, pp 327-330.

Van Harreveld A (1959) Compounds in brain extracts causing spreading depression of cerebral cortical activity and contraction of crustacean muscle. J Neurochem 3: 300-315

Van Harreveld A (1984) The nature of the chick's magnesium-sensitive retinal spreading depression. J Neurobiol 15: 333-344

Van Harreveld A, Fifková E (1970) Glutamate release from the retina during spreading depression. J Neurobiol 2: 13-29

Van Harreveld A, Fifková E (1971) Effects of glutamate and other amino acids on the retina. J Neurochem 18: 2145-2154

Wahl M, Lauritzen M, Schilling L (1987) The influence of cortical spreading depression upon pial arterial reactions to various vasoactive stimuli in cats and rats. Brain Res 411: 72-80

Welch KMA, Levine SR, D'Andrea G, Schultz LR, Helpern JA (1989) Preliminary observations on brain energy metabolism in migraine studied by in vivo phosphorus 31 NMR spectroscopy. Neurology 39: 538-541

Wolff HG (1963) Headache and other head pain. Oxford University Press, Oxford [illegible] (1987) [illegible] magnetic resonance [illegible] hippocampus [illegible] striatum. J Neuropath [illegible] 46: 380-395

[illegible] (1987) [illegible] for the primary site of [illegible] 15: 251

[illegible] (198?) [illegible] associated with spreading depression [illegible] Brain Res 4?? (1985) [illegible]

[illegible] (1990) Observations of [illegible] using [illegible] Wolfgang [illegible] (eds) [illegible]

[illegible] (19??) On [illegible] spreading depression [illegible] cerebral [illegible] cortex in [illegible] Neurophysiol 9: 40-415

Van Harreveld [illegible] magnetic resonance [illegible]

Van Harreveld [illegible] during spreading depression [illegible]

Van Harreveld [illegible] phosphate and [illegible]

Van Harreveld [illegible] (1987) [illegible] spreading depression [illegible] Brain Res 41: 75-87

Wisner [illegible] (1989) [illegible] spectroscopy [illegible] NMR [illegible]

Wu [illegible] (19??) [illegible] spreading depression [illegible]

GABAergic Mechanisms Involved in Cortical Spreading Depression in Normal and Early Malnourished Rats*

R. C. A. Guedes, J. E. Cabral-Filho, and N. R. Teodosio

Departamento de Nutrição, Universidade Federal de Pernambuco,
50739, Cidade Universitária, Recife, PE, Brasil

Introduction

Malnutrition early in life is known to produce remarkable alterations in structure and biochemistry of the nervous system, as well as in the behavior of adult animals (Gambetti et al. 1974; Cordero et al. 1976; Franková and Barnes 1968; Morgane et al. 1978; Brioni and Orsingher 1988). The electrophysiological activity is also significantly changed in malnourished animalls, both in the peripheral (Sima and Jankowska 1975; Silva et al. 1987) and in the central nervous system (Gramsbergen 1976; Morgane et al. 1978; Soto-Moyano et al. 1987). Also, there are reports of a greater susceptibility of malnourished rats to seizures (Stern et al. 1974; Taber et al. 1980) and to spreading depression (SD) (De Luca et al. 1977; Guedes 1984; Guedes et al. 1987a), two processes related to brain excitability.

The phenomenon of spreading depression SD has been described as a reversible and propagated "wave" of depression of the cortical spontaneous and evoked activity, with a simultaneous slow potential change of the cortical surface (Leão 1944, 1947; for a review, see Bureš et al. 1974). When elicited at one point of the cortical surface, by chemical (KCl), mechanical or electrical stimulation, SD spreads concentrically at a velocity of 2-5 mm/min (Leão 1944; Guedes and do Carmo 1980). The incidence and propagation of SD is known to be altered by some local and systemic conditions that alter neuronal excitability (Leão 1963; Martins-Ferreira et al. 1974; Bureš et al. 1974; Guedes and do Carmo 1980; Guedes 1984). This observation led to the idea of processes that were likely to be common to both phenomena (Leão 1944, 1972; Vieira-Neto et al. 1981; Guedes 1984).

A mechanism based on the neuronal release of chemical mediators has been postulated to be involved in SD propagation. Potassium (Grafstein 1956) and glutamate (Van Harreveld and Fifková 1970) are the two candidates most commonly discussed (see also Siesjo and Bengtsson 1989). Recently, the use of agonists and antagonists of neurotransmitters has

*Research was supported by Conselho Nacional de Desenvolvimento Científico e Tecnológico (CNPq) (No. 40.1991/84) and Financiadora de Estudos e Projetos (FINEP) (No. 43.890525.00). R.C.A.G. is the recipient of a CNPq fellowship (No. 30.1486/81).

This article is dedicated to Prof. Hiss Martins-Ferreira, on the occasion of his 70th birthday.

provided more direct evidence for the involvement of excitatory amino acid (Gorelova et al. 1987; Guedes et al. 1988) and opioid (Guedes et al. 1987b) mediated mechanisms in the genesis and propagation of SD.

GABAergic transmission is known to play an important role in the mechanism of seizures (Gale 1989; Peeters et al. 1989; Brailowsky et al. 1988; Le Gal La Salle et al. 1988), as well as in other neural functions (Meldrum 1982; Roberts 1986; Nicoll et al. 1990). There is less information for SD. Ochs and Hunt (1960) discussed the possibility of GABA suppressing cortical SD. Based on indirect evidence, a possible inhibitory role of GABA on SD in the olfactory bulb has been suggested by Amemori et al. (1987). Rodrigues et al. (1988) have shown a significant release of GABA from the isolated chicken retina during SD. Preliminary results from our laboratory have demonstrated the ability of bicuculline to evoke SD (Guedes 1984). All these data suggest the participation of GABA-mediated mechanisms in SD. The present work was undertaken to investigate this possibility in the rat cerebral cortex, by using the GABA agonist diazepam (DZP) and the antagonists bicuculline methiodide (BMI) and picrotoxin (PTX). As the GABAergic transmission (Telang et al. 1984; Almeida et al. 1988; Brioni and Orsingher 1988; Brioni et al. 1989) and SD susceptibility (De Luca et al. 1977; Guedes 1984; Guedes et al. 1987a) have been shown to be altered in early malnourished rats, it was considered also of interest to examine the GABA-mediated processes probably involved in SD in adult rats submitted to early malnutrition.

Methods

Early protein deficiency (EPD) was provoked in 31 Sprague-Dawley rats during the suckling period by feeding their mothers a low-protein diet (8% protein from defatted, powdered milk). All other ingredients of the diet were kept at normal levels. After weaning the pups were fed a control diet (22% protein) until adult life (90-150 days), when the recording sessions were performed. These EPD rats were compared with 28 controls, tested at the same age range. Three groups of EPD and three groups of control animals were formed. Surgical procedures for the recording sessions were performed with the rats under anesthesia with urethane-chloralose mixture (750 mg/kg + 60 mg/kg, respectively, i.p.). A catheter was inserted into one femoral vein and three trephine holes were drilled on one side of the skull; the more frontal hole was used for KCl stimulation, to evoke SD, and through the other two holes both the electrocorticogram (ECoG), as well as the recording of the slow potential change (P) accompanying SD (Leão 1944, 1947) were performed. SD was elicited by 1 min application of KCl (2% solution) in a small pledget of cotton (1 mm diameter), placed directly on the surface of the cortex. Recordings were made continuously, for a period of at least 6h, by means of cotton wick, and Ag-AgCl agar-Ringer electrodes, as previously described (Guedes and do Carmo 1980). Two such electrodes were placed in contact with the cortical surface (one through each recording hole) and a common reference electrode was located on the nasal bones. The interval between successive KCl stimuli was at least 20 min. Rectal temperature was maintained at 37.5° ± 1°C by means of a heating blanket. The velocity of SD propagation was calculated based on the interelectrode distance, checked several times during the recording period, and the time necessary for an SD wave to cross this distance. In each animal, SD velocity was measured several times before and after injecting a given drug (only one drug per animal was tested). Each drug was tested in one control and one EPD group. Drugs injected intravenously during the recording sessions were: DZP (1.5, 4.0, and 8.0 mg/kg; 12 control and 7 EPD rats), BMI (0.15, 0.6, 1.5, 3.0, and 6.0 mg/kg; 8 control and 10 EPD rats), and PTX (1.0, 1.5, 2.0, and 4.0 mg/kg; 8 control and 14 EPD rats). At the end of the recording sessions the animals were killed with an overdose of anesthesic and the brains removed immediately and weighed (wet weight). They were then kept at 100°C and weighed daily until a constant weight was attained (dry weight). Two-tailed *t* tests were used for comparisons (paired for dependent samples and not paired for independent ones).

Results

Effects of EPD on Body and Brain Weights

In adult rats that were subjected to EPD during the suckling period, mean body and brain weights were significantly lower than those of the control animals. Body weight of EPD rats was 49.4% that of controls, while both wet and dry brain weights were 86.3% those of controls. Data on body and brain weights are shown in Table 1. The difference in the body weight between males and females, usually observed in control rats, was not present in the EPD groups, where males and females had similar body weights (147.5 ± 30.5 and 136.8 ± 15.3 g, respectively).

Table 1. Body and brain weights and velocity of SD propagation (pre-drug condition) in control and early malnourished adult rats

Groups	Weights			SD velocity (mm/min)					
		Brain (mg)		Hour					
	Body (g)	Wet	Dry	1	2	3	4	5	6
Control	282.3	1666.6	302.7	3.43	3.45	3.66	3.72	3.83	3.87
	± 66.6	± 82.0	± 37.0	± 0.33	± 0.51	± 0.72	± 0.74	± 0.72	± 0.81
	(26)	(9)	(8)	(25)	(19)	(15)	(13)	(10)	(8)
Early Malnourished	139.5*	1437.8*	261.1*	4.02*	4.26*	4.44*	4.89*	4.63*	4.75*
	± 19.9	± 88.0	± 12.0	± 0.73	± 0.94	± 1.01	± 0.84	± 0.56	± 0.83
	(24)	(10)	(10)	(31)	(26)	(24)	(16)	(14)	(13)

Data are reported as means ± SD with the number of measurements given in parentheses.
* significantly different from the respective control, $p < 0.05$, *t* test

Effects of EPD on SD Velocity

The values for SD velocity obtained in the EPD groups before drug injections were compared with the values obtained in the control animals in the same situation. As can be seen in Table 1, EPD rats displayed SD velocities significantly higher than those of the controls, confirming previous reports showing a facilitatory effect of early malnutrition on SD propagation in adult rats.

Effects of DZP, BMI, and PTX in Control Rats

In 12 control animals, the first i.v. injection of DZP (performed 3-6 min before SD elicitation) resulted in a significant increase (9%; $p < 0.01$) in the velocity of SD propagation (Fig. 1). A second injection in the same animal produced less evident effects, leading to a decrease in SD velocity in some animals and to an increase in others, so that the mean difference was not significant. When three or more DZP injections were performed in the same animal within a period of 60 min (three rats), the ECoG activity became very depressed, SD velocity decreased markedly, and two animals died.

SD velocity measured 5-30 min after BMI injection was reduced by 8%, which was statistically significant ($p < 0.01$) compared with the pre-injection value (Fig. 1). A second injection of BMI did not have a significant additional effect on SD velocity, although in most animals (five out eight rats tested) there was a tendency toward lower SD velocities compared with the values before this second injection.

The first injection of PTX led to the appearance of seizure activity and caused a significant reduction (15%, $p < 0.01$) in velocity of SD propagation (Fig. 1). The mean latency for the occurrence of this effect was greater than for the effect of BMI (5 min and 15 min, respectively), so that for PTX we used a longer interval (10-40 min) between injection and SD initiation. Like the other two drugs (BMI and DZP), a second injection of PTX did not have a significant additional effect on the SD velocity.

Conspicuous ECoG changes induced by BMI and PTX were characterized by the appearance of seizure activity with large-amplitude, spike-like waves which, in some cases, triggered SD. Clonic movements of various intensities could sometimes be observed to occur in a dose-related fashion, after the administration of either convulsant. In order to test for a possible interference of these motor concomitants in the elicitation of SD, as well as in the SD velocity, some animals were paralyzed with gallamine (40 mg/kg i.v.) and artificially ventilated before BMI or PTX was injected. No difference could be detected between these animals and the unparalyzed ones, with regard to the effect of the convulsants on velocity of SD propagation, or on the elicitation of SD. By increasing the amount of either convulsant we were able to produce a prolonged status of intense seizure-like EEG activity. At this time, SD could no longer be elicited by KCl stimulation (Fig. 2). As the seizure activity gradually decreased (30-60 min after having reached its maximum), KCl stimulation again

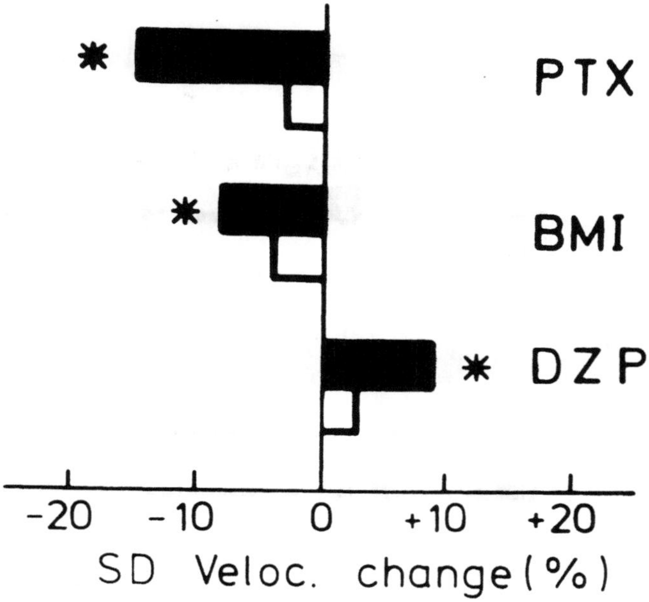

Fig. 1. Mean change in velocity of SD in rats treated with i.v. 1.0-4.0 mg/kg picrotoxin (*PTX*), 0.15-6.0 mg/kg bicuculline (*BMI*), and 1.5-8.0 mg/kg diazepam (*DZP*) in well-nourished (*filled bars*) and early protein deficient (*open bars*) animals. Results are expressed as percentages of the velocity before drug injection. * $p < 0.01$, paired t test

became effective in eliciting SD. Examples of recordings made before and after injections of the three drugs are shown in Fig. 2.

Effects of DZP, BMI, and PTX in EPD Rats

In the EPD groups, the effects of the three drugs on the ECoG, as well as on the velocity of SD propagation, were less conspicuous. None of the three drugs produced significant alterations in the velocities of SD propagation even after a second injection. When expressed as a percentage of the preinjection values, SD velocities were 103% (DZP), 96% (BMI), and 97% (PTX), not significantly different from the respective controls (Fig. 1).

Discussion

The participation of GABAergic transmission in seizure generation has been documented in previous studies (see Gale 1989). GABA is also suggested to have an inhibitory action on SD in the cerebral cortex of rabbits (Ochs and Hunt 1960) and in the rat's olfactory bulb (Amemori et al. 1987). In the present experiments, P and ECoG activity during SD were measured and DZP-, BMI-, and PTX-induced alterations in these activities, as well as in the velocity of SD propagation, were also monitored. The data suggest a facilitatory participation of the GABA system in SD propagation. This seems surprising in view of the abovementioned suggestions by Ochs and Hunt (1960) and Amemori et al. (1987), but it is in

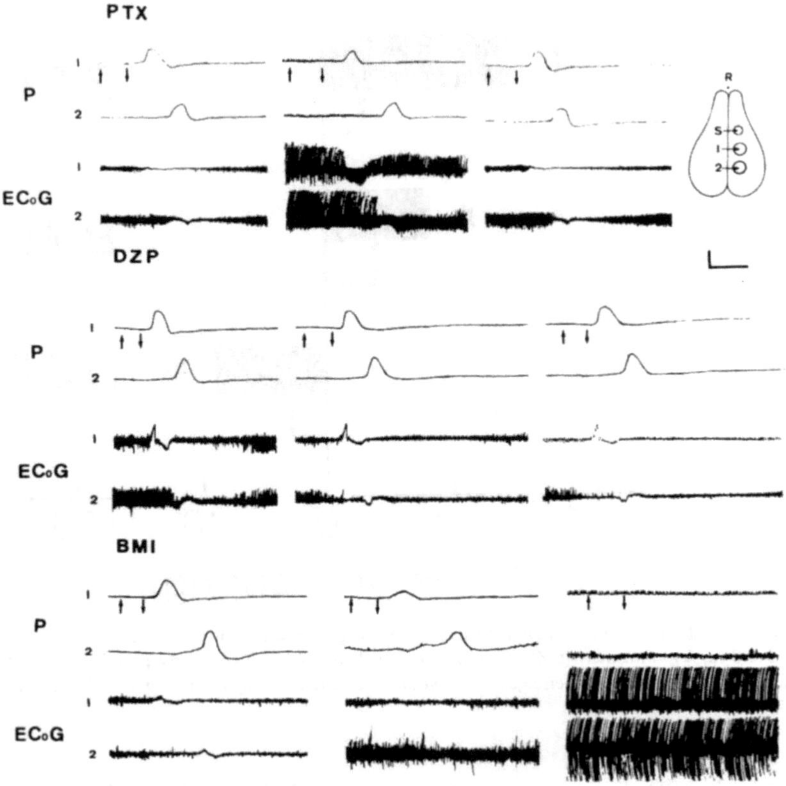

Fig. 2. Effects of i.v. injection of picrotoxin (*PTX*) diazepam (*DZP*) and bicuculline (*BMI*) on the electrocorticogram (*ECoG*) and SD initiation in well-nourished rats. Sections taken from continuous recording of three experiments. *Left column*, control. *Middle and right columns*, after drug injection. *Arrows* denote the period of topical administration of 2% KCl solution to elicit SD. Test stimulation was applied 19 and 60 min after 1.0 mg/kg PTX (*upper panel*); 6 and 50 min after 5.0 mg/kg DZP (*middle panel*); 24 min after the injection of 0.6 mg/kg BMI and 115 min after a second injection of the drug (1.5 mg/kg) which was administered 50 min after the first (*bottom panel*). Note the changes in latency after the administration of the drugs (middle column), effects that were reversed in the second test (*right column*) for PTX and DZP. In the *bottom panel* (*right*) is seen the intense convulsive activity produced by the second injection of BMI. Under this condition, stimulation was ineffective at eliciting SD. *Inset*, position of the cortical (*1,2*) and reference (*R*) electrodes, and the stimulation site (*S*). *Calibration bars*, horizontal, 1 min; vertical, 10 mV for the slow potential change (*P*), and 0.3 mV for ECoG

agreement with the reports of Rodrigues et al. (1982), who showed that GABA is able to start SD in the retina. Also, Martins-Ferreira et al. (1984) have demonstrated that microinjections of GABA in the retina cause the release of potassium in concentrations high enough to trigger SD. This raises the question of how the GABA system might influence or participate in the phenomenon of SD.

One of the reasons for the suggestion of an inhibitory role of GABA on SD in the assumption that the facilitation of neural excitability (Ramson 1974; Chamberlin and Dingledine 1988) and of SD (Leão 1963; Martins-Ferreira et al. 1974; Guedes and do Carmo 1980), which is seen in low-chloride conditions, is due to the blockade of GABAergic activity associated with Cl^- flow through GABA-operated channels (Amemori et al. 1987). However, this suggestion is based on indirect inferences regarding GABAergic mechanisms at the synaptic level and, as such, may not consider more specific or selective mechanisms involved in the phenomenon.

In fact, our results with BMI, a GABA antagonist that competes specifically at the receptor, indicate that direct blockade of the receptor may be responsible for the reduction in velocity of SD propagation. Furthermore, the results with drugs acting in a more selective manner — PTX blocking and DZP facilitating the flow in the Cl^- channel associated with the GABA-benzodiazepine receptor complex (Braestrup and Nielsen 1980; Olsen et al. 1984) — also indicate an involvement of the GABA system. It is possible, however, that the action of this system in SD is double, depending upon which parameter is considered. In fact, we also observed that i.v. DZP, besides an increase in SD velocity, often produced a reduction in P amplitude. These modulatory effects of the GABA system suggest the existence of different processes in the brain tissue underlying the P amplitude and the SD propagation. This possibility had already been discussed by Bureš et al. (1984).

It must be pointed out that the present results and those of others indicating the influence of putative neurotransmitters on SD (Van Harreveld and Fifková 1970) do not provide definitive evidence of the involvement of synaptic mechanisms in the phenomenon. Synaptic activity may play a role in the initiation of SD but does not seem to be required for its propagation, since SD spreads without difficulty from a normal cortical area to another in which action potentials were abolished by tetrodotoxin (Sugaya et al. 1975). Also we cannot conclude that the GABA system is the only one involved in SD, because of the interactions between this and other systems (Nutt and Cowen, 1987). In addition, SD has been shown to be affected by means of the manipulation of the excitatory amino acid system (Gorelova et al. 1987; Guedes et al. 1988) as well as of the opioid system (Sprick et al. 1981; Guedes et al. 1987b). More detailed studies will be necessary to clarify the SD-transmitter interaction.

Our data confirm previous reports from this and other laboratories, concerning the effects of malnutrition on body and brain weights. Similarly, the effects on SD velocity of propagation are in agreement with comparable results from other models of malnutrition (De Luca et al. 1977; Guedes 1984; Guedes et al. 1987a). GABAergic activity has been shown to be diminished in malnourished animals (Telang et al. 1984; Wiggins et al. 1984). A reduction of the anticonflict effect of DZP (Brioni et al. 1989) and of chlordiazepoxide (Almeida et al. 1988) in malnourished rats has also been reported. Accordingly, in the present work DZP, BMI and PTX affected SD propagation in normal rats and failed to do so in malnourished rats. Since the number of GABA receptors does not seem to be reduced in malnutrition (Telang et al. 1984), the lack of drug effect on SD could be due either to a reduced receptor affinity or to a diminished amount of GABA available to interact with it. Our data do not permit a choice between these two possibilities, but they clearly show an impairment of the brain GABAergic function produced by malnutrition. Finally, it must be kept in mind that malnutrition can affect different neural structures and cell types differently and this may help us to understand some of the conflicting data in the literature regarding the effects of malnutrition on the morphological, biochemical, and functional development of the nervous system.

Acknowledgements. We are very grateful to Ana Maria de França Silva and Hamilton Salviano da Silva for technical assistance, to Fernanda Maria de Almeida for secretarial help, and to Produtos Roche Químicos e Farmacêuticos S/A for donating the vitamin mixture for the diets.

Summary

The participation of γ-aminobutyric acid (GABA)-mediated mechanisms in the phenomenon of cortical spreading depression (SD) was investigated in well-nourished (control) and early malnourished adult rats injected intravenously during the recording session with diazepam (DZP), bicuculline methiodide (BMI), or picrotoxin (PTX).

In control animals, the three drugs significantly altered the velocity of SD propagation: DZP increased it (9%), and BMI and PTX decreased it (8% and 15%, respectively). Before administration of the drugs, early malnourished rats presented mean SD velocities significantly higher than their controls, confirming previous reports on other models of malnutrition. In early malnourished animals, SD velocity was not altered by drug injections.

These results suggest a facilitatory participation of the GABA system in SD propagation. Malnutrition, on the other hand, seems to abolish this effect.

References

Almeida SS, De-Oliveira LM, Bichuette MZ, Graeff FG (1988) Early malnutrition alters the effect of chlordiazepoxide on inhibitory avoidance. Braz J Med Biol Res 21: 1033-1036

Amemori T, Gorelova NA, Bureš J (1987) Spreading depression in the olfactory bulb of rats: reliable initiation and boundaries of propagation. Neuroscience 22: 29-36

Braestrup C, Nielsen M (1980) Benzodiazepine receptors. Arzneimittelforsch 30: 852-857

Brailowsky S, Kunimoto M, Menini C, Silva-Barat C, Riche D, Naquet R (1988) The GABA-withdrawal syndrome: a new model of focal epileptogenesis. Brain Res 442: 175-179

Brioni JD, Orsingher OA (1988) Operant behavior and reactivity to the anticonflict effect of diazepam in perinatally undernourished rats. Physiol Behav 44: 193-198

Brioni JD, Cordoba N, Orsingher OA (1989) Decreased reactivity to the anticonflict effect of diazepam in perinatally undernourished rats. Behav Brain Res 34: 159-162

Bureš J, Burešová O, Křivánek J (1974) The mechanism and applications of Leão's spreading depression of electroencephalographic activity. Academia, Prague

Bureš J, Burešová O, Křivánek J (1984) The meaning and significance of Leão's spreading depression. An Acad Bras Ciênc 56: 385-400

Chamberlin NL, Dingledine R (1988) GABAergic inhibition and the induction of spontaneous epileptiform activity by low chloride and high potassium in hippocampal slice. Brain Res 445: 12-18

Cordero ME, Diaz G, Araya J (1976) Neocortex development during severe malnutrition in the rat. Am J Clin Nutr 29: 358-365

De Luca B, Cioffi LA, Bureš J (1977) Cortical and a caudate spreading depression as an indicator of neural changes induced by early malnutrition in rats. Act Nerv Super 19: 130-131

Franková S, Barnes RH (1968) Effect of malnutrition in early life on avoidance conditioning and behavior of adult rats. J Nutr 96: 485-493

Gale K (1989) GABA in epilepsy: the pharmacologic basis. Epilepsia 30 [Suppl 3]: S1-S11

Gambetti P, Gambetti LA, Rizzuto N, Shafer B, Pfaff L (1974) Synapses and malnutrition: quantitative ultrastructural study of rat cerebral cortex. Exp Neurol 43: 464-473

Gorelova NA, Koroleva VI, Amemori T, Bureš J (1987) Ketamine blockade of cortical spreading depression in rats. Electroencephalogr Clin Neurophysiol 66: 440-447

Grafstein B (1956) Mechanism of spreading cortical depression. J Neurophysiol 19: 154-171

Gramsbergen A (1976) EEG development in normal and undernourished rats. Brain Res 105: 287-308

Guedes RCA (1984) On some conditions that influence cortical spreading depression. An Acad Bras Ciênc 56: 445-455

Guedes RCA, do Carmo RJ (1980) Influence of ionic disturbances produced by gastric washing on cortical spreading depression. Exp Brain Res 39: 341-349

Guedes RCA, Andrade AFD, Cabral-Filho JE (1987a) Propagation of cortical spreading depression in malnourished rats: facilitatory effect of dietary protein deficiency. Braz J Med Biol Res, 20: 639-642

Guedes RCA, Azeredo FAM, Hicks TP, Clarke RJ, Tashiro T (1987b) Opioid mechanisms involved in the slow potential change and neuronal refractoriness during cortical spreading depression. Exp Brain Res 69: 113-119

Guedes RCA, Andrade AFD, Cavalheiro EA (1988) Excitatory amino acids and cortical spreading depression. In: Cavalheiro EA, Lehman J, Turski L (eds) Frontiers in excitatory amino acid research. Liss, New York, pp 667-670

Leão AAP (1944) Spreading depression of activity in the cerebral cortex. J Neurophysiol 7: 359-390

Leão AAP (1947) Further observations on spreading depression of activity in the cerebral cortex. J Neurophysiol 10: 409-414

Leão AAP (1963) On the spread of spreading depression. In: Brazier MAB (ed) Brain function. University of California Press, Berkeley, pp 78-85

Leão AAP (1972) Spreading depression. In: Purpura DP, Penry K, Tower DB, Woodbury DM, Walter RD (eds) Experimental models of epilepsy. Raven, New York, pp 173-195

Le Gal La Salle G, Brailowsky S, Menini C, Naquet R (1988) Local asymptomatic status epilepticus induced by withdrawal of GABA infusion into limbic structures. Exp Neurol 101: 411-417

Martins-Ferreira H, Oliveira-Castro G, Struchiner CJ, Rodrigues PS (1974) Circling spreading depression in isolated chick retina. J Neurophysiol 37: 773-784

Martins-Ferreira H, do Carmo RJ, Netto M (1984) Potassium release by spreading depression evoking substances. An Acad Bras Ciênc 56: 116-117

Meldrum B (1982) Pharmacology of GABA. Clin Neuropharmacol 5: 293-316

Morgane PJ, Miller M, Kemper T, Stern W, Forber W, Hall R, Bronzino J, Kissane J, Hawrylewicz E, Resnick O (1978) The effects of protein malnutrition on the developing central nervous system in the rat. Neurosci Biobehav Rev 2: 137-230

Nicoll RA, Malenka RC, Kauer JA (1990) Functional comparison of neurotransmitter receptor subtypes in mammalian central nervous system. Physiol Rev 70: 513-565

Nutt DJ, Cowen PJ (1987) Diazepam alters brain 5-Ht function in man: implications for the acute and chronic effects of benzodiazepines. Psychol Med 17: 601-607

Ochs S, Hunt K (1960) Apical dendrites and propagation of spreading depression in cerebral cortex. J Neurophysiol 23: 432-444

Olsen RW, Wong EHF, Stauber GB, King RG (1984) Biochemical pharmacology of the γ-aminobutyric acid receptor/ionophore protein. Fed Proc 43: 2773-2778

Peeters BWMM, Van Rijn CM, Vossen JMH, Coenen AML (1989) Effects of GABA-ergic agents on spontaneous non-convulsive epilepsy, EEG and behaviour, in the Wag/Rij inbred strain of rats. Life Sci 45: 1171-1176

Ramson BR (1974) The behavior of presumed glial cells during seizure discharge in cat cerebral cortex. Brain Res 69: 83-99

Roberts E (1986) What do GABA neurons really do? They make possible variability generation in relation to demand. Exp Neurol 93: 279-290

Rodrigues PS, Azeredo FAM, De Mello FG, Martins-Ferreira H (1982) Does gamma-aminobutyric acid (GABA) participate in retinal spreading depression (SD)? Neuroscience [Suppl] 7: 179

Rodrigues PS, Guimarães APO, de Azeredo FAM, Martins-Ferreira H (1988) Involvement of GABA and ACh in retinal spreading depression: effects of "low calcium-high magnesium" solutions. Exp Brain Res 73: 659-664

Siesjo BK, Bengtsson F (1989) Calcium fluxes, calcium antagonists and calcium-related pathology in brain ischemia, hypoglycemia and spreading depression: a unifying hypothesis. J Cereb Blood Flow Metab 9: 127-140

Silva AT, da Costa FBR, Costa JA, Teodósio NR, Cabral Filho JE, Guedes RCA (1987) Sciatic nerve conduction velocity of malnourished rats fed the human "Basic Regional Diet" of the northeast of Brazil. Braz J Med Biol Res 20: 383-392

Sima A, Jankowska E (1975) Sensory nerve conduction velocity as correlated to fibre size in experimental undernutrition in the rat. Neuropathol Appl Neurobiol, 1: 31-37

Soto-Moyano R, Ruiz S, Pérez H, Hernandez A, Belmar J (1987) Early undernutrition and long-lasting functional derangement of the noradrenergic system projecting to the neocortex. Nutr Rep Int 36: 309-317

Sprick U, Oitzl MS, Ornstein K, Huston JP (1981) Spreading depression induced by microinjection of enkephalins into the hippocampus and neocortex. Brain Res 210: 243-252

Stern WC, Forbes WB, Resnick O, Morgane PJ (1974) Seizure susceptibility and brain amine levels following protein malnutrition during development in the rat. Brain Res 79: 375-384

Sugaya E, Takato M, Noda Y (1975) Neuronal and glial activity during spreading depression in cerebral cortex of cat. J Neurophysiol 34: 822-841

Taber KH, Fuller GN, Stanley JND, France JF, Wiggins RC (1980) The effect of postnatal undernourishment on epileptiform kindling of dorsal hippocampus. Experientia 36: 69-70

Telang S, Fuller G, Wiggins R, Enna SJ (1984) Early undernutrition and (H^3)-aminobutyric acid binding in rat brain. J Neurochem 640-645

Van Harreveld A, Fifková E (1970) Glutamate release from the retina during spreading depression. J Neurobiol 2: 13-29

Vieira-Neto A, Guedes RCA, Teodósio NR (1981) Lasting effects of acute dehydration and post-weaning undernourishment on cortical spreading depression in adult rats. Experientia 37: 738-740

Wiggins RC, Fuller G, Enna SJ (1984) Undernutrition and the development of brain neurotransmitter systems. Life Sci 35: 2085-2094

Spreading Depression and Neuron Damage: a Brief Review *

G. G. Somjen, O. Herreras, and J. Jing

Division of Physiology, Duke University Medical Center, Durham, NC 27710, USA

Introduction

Almost 50 years after its discovery, *depressão alastrante* of Leão continues to puzzle and to intrigue. Studied intensively in numerous laboratories around the world, its mechanism is still poorly understood. Is it part of the normal function of the brain that has evolved for some specific as yet undisclosed purpose, or it is strictly linked to pathology? If it is pathophysiological, is it a defensive response or is it part of the noxious process? What can trigger it outside the laboratory, in "real life"? What are the ingredients of the positive feedback loops that bring it about and what are the factors that ordinarily limit its duration? And does it have lasting aftereffects?

In this brief review we address the last question. There are at least three conditions in human pathology where spreading depression (SD), or a process closely related to it, are likely to play a part. These are: hypoxia or ischemia of brain tissue, cerebral concussion, and migraine (for reviews, see: Lauritzen 1987; Somjen 1988, 1991). Apart from any theoretical interest, for the management of these conditions it would be important to know what the profound depolarization of cells and the redistribution of ions during SD mean for the health of the brain.

The similarity of the extracellular potential shift associated with SD and that seen when the blood flow to the cortex is interrupted was first reported by Leão in 1947. Whether or not the two phenomena, SD of normoxic gray matter and the SD-like event of ischemia, are identical is still unresolved. Marshall (1959), for one, believed they may be the same, but others had doubts. Bureš and Burešová (1957) suspected that the sudden, profound depolarization of cells in ischemic cerebral cortex heralds the onset of irreversible damage, and they coined the term "terminal depolarization" to describe the event.

*Supported by grants NS 17771 and NS 18670 of the NIH, USPHS, and grant PC89 to O.H. from the Spanish Ministry of Education.

Hypoxic SD and Neuron Damage

SD-like hypoxic depolarization need not be lethal, if reversed in time. When the supply either of blood or of oxygen to cerebral tissue is interrupted, whether in situ or in vitro, first spontaneous EEG activity ceases, then all synaptic transmission fails, and only some minutes later does SD-like depolarization set in. If oxygenation or perfusion with blood is restored promptly, cerebral function recovers rapidly. If, however, the condition persists beyond a critical time, then synaptic function remains depressed in spite of belated reoxygenation (e.g., Balestrino and Somjen 1986; Balestrino et al. 1989). The question is whether the depolarized state itself plays a part in irreversible depression of function and eventual cell loss, or whether it is merely coincidental to other pathologic processes. We find both theoretical reasons and experimental observations which suggest that protracted deep depolarization is in fact bad for neurons.

It is known that during SD as well as hypoxic SD-like depolarization calcium ions move from extracellular into intracellular fluid (Somjen and Aitken 1984; Hansen 1985). Excessive elevation of the concentration of free calcium in the cytoplasm is believed to trigger a chain of biochemical reactions which ultimately leads to the self-destruction of cells (Siesjö 1981). We noted earlier, however, that neurons can recover fully from brief hypoxic SD. Critical for cell damage is apparently not just the elevation of intracellular free calcium, but the length of time during which the condition persists (Balestrino and Somjen 1986; Balestrino et al. 1989; Somjen 1991; Somjen et al. 1990).

Recurrent SD in Normoxic Brain Tissue

Calcium accumulates in cells during SD in normoxic cerebral tissue and, if this causes cell damage, then damage should be evident after normoxic SD as well as in ischemic/hypoxic brain. Yet many waves of SD can pass without evidence of lasting harm (Ruščáková 1964; Burešová and Bureš 1969; Nedergaard and Hansen 1988). But in these experiments SD came in waves lasting about a minute with longer rest periods in between, and it may be that during the repolarized intervals the hypothetical destructive biochemical processes were reversed. We therefore designed methods to test the effect of prolonged, continuous, profound depolarization and recovery of function.

Protracted SD-Like Depolarization of Neurons in Hippocampal Tissue Slices

Kawasaki et al. (1988) irrigated the surface of hippocampal tissue slices with a solution containing a high potassium concentration while recording interstitial K^+ concentration ($[K^+]_o$) with ion-selective electrodes. Exposure to high K^+ resulted in SD-like shift of extracellular voltage and sudden elevation of $[K^+]_o$. The SD-like state was maintained as long as the slice was irrigated with high K^+, and subsided over several minutes thereafter. Synaptic transmission recovered completely if the SD-like depolarization lasted 4 min or less. With SD lasting 9 min or more there was an initial partial recovery followed by secondary failure of the orthodromically transmitted population spike, which did not recover within the 1-3 h of observation. The focally recorded excitatory postsynaptic potential (fEPSP) was, however, partially preserved.

The observations of Kawasaki et al. (1988) agreed with the notion that normoxic prolonged SD-like depolarization causes lasting cell damage, but they were incomplete. The period allowed for recovery after high K^+ administration was only 1-3 h, and in a longer period function could perhaps have been restored more completely. Moreover, antidromic conduction was not tested. The cells may have been viable, albeit unable to transmit orthodromically. Currently we are continuing this study. While data are as yet incomplete, a few facts have emerged.

We have exposed hippocampal slices to elevated K^+ for 10 min, and followed the course of recovery for 5-6 h thereafter. While in most cases synaptic transmission remained severely depressed, in four out of ten slices the orthodromic population spike recovered 60%-100% of its control amplitude. Even in the more severely depressed slices, antidromic population spikes regained somewhat more of amplitude than did orthodromic spikes. We concluded that a depressed state 1 h after an insult does not necessarily signify irreversible neuron damage.

Functional Recovery After Recurrent and After Protracted SD in Hippocampus of Anesthetized Rats

When brain tissue slices are maintained in vitro, conditions are far from clinical reality. For this reason we have begun testing the effect of prolonged SD-like states on hippocampus of anesthetized rats. This work is also incomplete, and our conclusions are tentative.

Spreading depression is induced in two ways. In one series of experiments a microdialysis fiber (220 μm OD, 800 μm exposed length) is inserted in CA1 region of dorsal hippocampus. Recordings are made from a microwire attached to the dialysis probe, and also from one or more extracellular micropipette electrodes, or a double-barreled K^+-selective microelectrode, inserted 0.5-1.2 mm rostral to the dialysis probe, in various layers of CA1 in different experiments. The DC voltage is recorded on chart paper, as well as potential responses evoked by stimulating CA3 region of the same side, and sometimes also (for antidromic activation) the alveus. When control recordings indicate a stable state of the preparation, artificial cerebrospinal fluid (ACSF), in which 100 m*M* K^+ has replaced Na^+, is perfused continuously for 30-90 min through the dialysis probe. Recovery of function is then observed for 3-5 h.

Figure 1 shows DC-coupled recordings made simultaneously from stratum pyramidale (cell bodies) and stratum radiatum (apical dendrites) approximately 0.8 mm from the dialysis fiber during high K^+ dialysis. Soon after the start of the high K^+ administration, a typical SD occurred, which recovered after about 1 min. Additional waves of SD followed, and then a state of instability developed in which the dendritic region remained at a strongly negative voltage for more than 7 min interrupted a few times by only brief, abortive deflections toward baseline potential. In this period stratum pyramidale showed only an irregular negativity of moderate amplitude. Provisionally we call this phenomenon "prolonged unstable SD." During the second half of the high K^+ administration several more severe and prolonged SD waves occurred and another, late, period of instability.

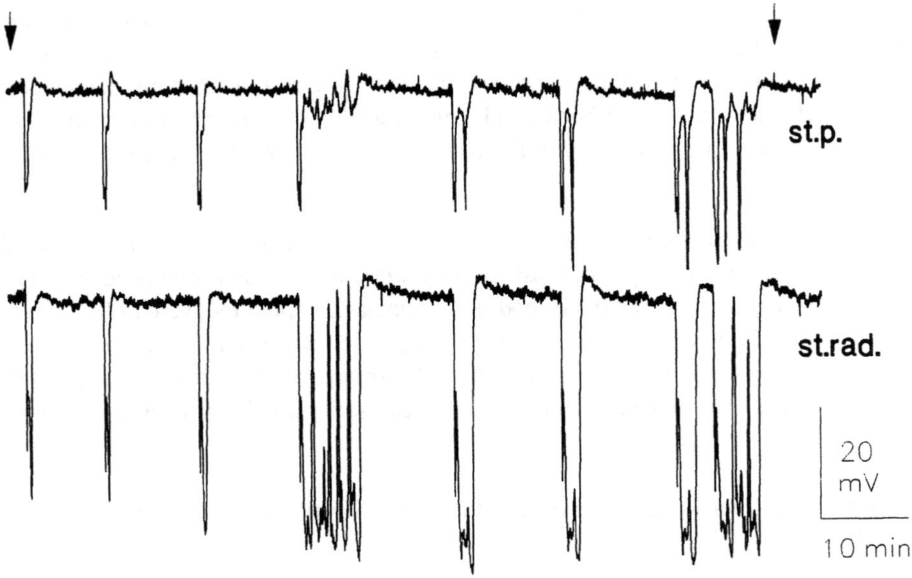

Fig. 1. Recurrent SD and prolonged unstable SD in CA1 of hippocampus. The *upper trace* was recorded in stratum pyramidale (cell body layer) and the *lower trace* in stratum radiatum (apical dendrites). *Between the arrows* ACSF with elevated (100 m*M*) K^+ was administered by microdialysis

All similarly treated hippocampi showed recurrent SD, with the waves of potential shifts growing through the initial three to six episodes. SD in stratum radiatum always had a double negative peak, and it was always the second component that grew with repetition. Prolonged unstable SD was seen in 11 out of 22 experiments. When $[K^+]_o$ was recorded, it was seen elevated throughout SD waves and throughout protracted instability in stratum radiatum, but the contour of the $[K^+]_o$ rise was not an exact mirror image of the potential shift. In stratum pyramidale, $[K^+]_o$ was high during SD, but only slightly elevated during prolonged SD. A steep dorsoventral gradient of $[K^+]_o$ appears to be maintained throughout the state of instability.

Evoked potentials were completely suppressed during SD waves and also during prolonged SD. Between SD episodes, during high K^+ perfusion, synaptic transmission partially recovered, often with signs of hyperexcitability such as multiple orthodromic population spike responses. Antidromic spikes usually recovered to normal amplitude between SD episodes.

After high K^+ dialysis, evoked potentials first recovered, and there were signs of hyperexcitability. In a number of instances this was followed by secondary depression of the responses. In cases when prolonged SD has not been induced during high K^+ dialysis, secondary

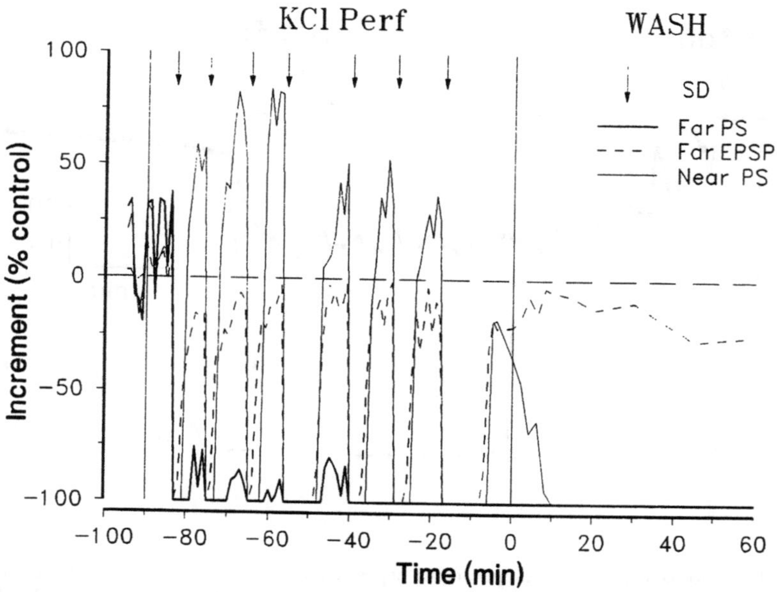

Fig. 2. Amplitude of evoked responses plotted against elapsed time. *Far PS* is orthodromically transmitted population spike recorded in stratum pyramidale 0.8 mm from dialysis fiber; *Far EPSP* is extracellular excitatory postsynaptic potential recorded simultaneously from stratum radiatum, 300 μm deeper but same AP and lateral coordinates as *far PS*; *Near PS* is population spike recorded from microwire attached to dialysis probe. Zero time is the end of high K^+ administration; zero amplitude equals control level recorded at beginning of session. *Vertical arrows* mark onset of SD waves (as illustrated in Fig. 1)

depression was followed by renewed recovery to normal amplitude in 90-300 min. In the brains which had undergone prolonged SD, recovery was incomplete: 5 h after washing out the high K^+, orthodromic spike amplitude varied from 16%-83% of control amplitude.

In other experiments the dorsal hippocampus was exposed by removing the overlying cortex. The exposed surface was irrigated by ACSF containing 100 m*M* K^+ and, additionally, the same solution was also perfused through a dialysis probe, as before. Recordings were made within 0.4 mm of the dialysis probe. In these cases a protracted stable negative shift indicated continuous depolarization of cells (Fig. 3). Orthodromic transmission recovered completely after 30 min of such treatment, partially after 45 min (Fig. 3), and it was lost for at least 5 h after 60 or more minutes of high K^+ administration (Fig. 4).

Conclusions

These newer experiments confirm that protracted, profound depolarization harms neurons, whereas recurrent, brief episodes of depolarization do not. We must not underestimate, however, the power of healing in nerve cells. Absence of synaptic transmission 1 h after an insult cannot be taken as proof of irreversible damage.

Comparing our recent observations with our earlier ones, and with those made by others (reviewed in Somjen 1988, 1991; Somjen et al. 1990), it appears that hypoxic SD causes

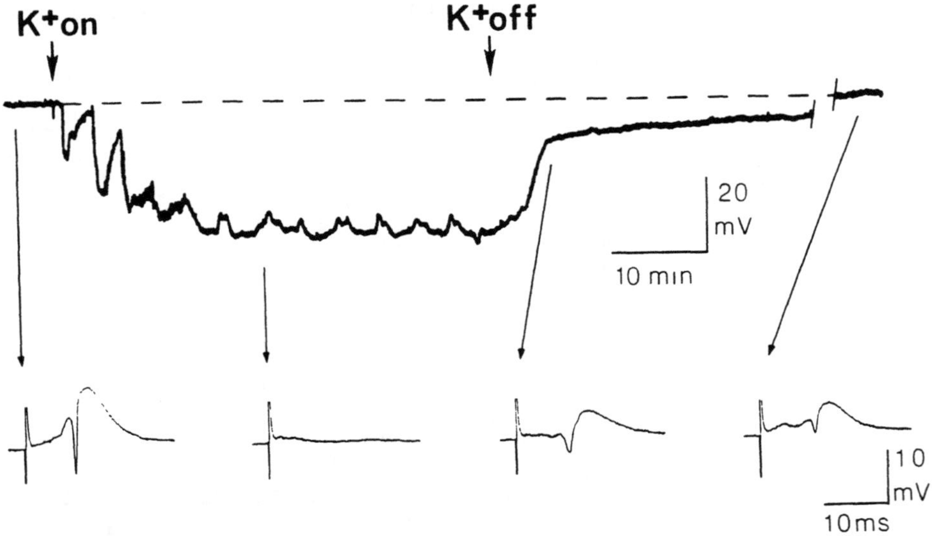

Fig. 3. Recordings from a hippocampus exposed for 45 min to elevated K^+ by irrigation of the surface plus microdialysis. *Upper tracing*, DC-coupled record from CA1 stratum pyramidale. *Lower row of records*: orthodromic evoked potentials. At the break the DC trace is interrupted for 280 min. *Thin arrows* mark times of evoked potential recordings

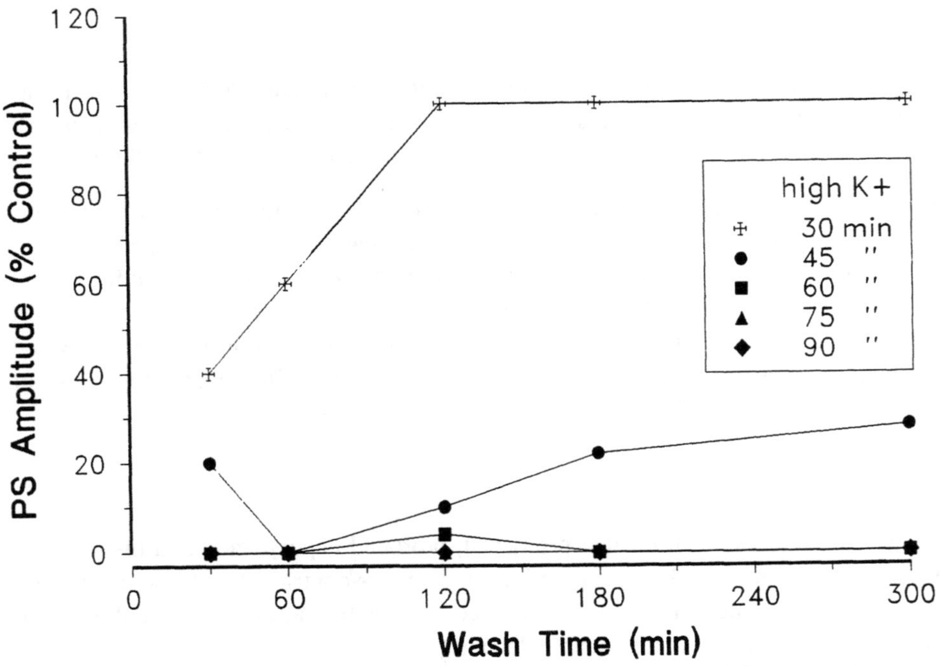

Fig. 4. Orthodromic population spike amplitude recorded various times after high K^+ administration. *Inset* shows duration of high K^+ exposure. *Points* for 30 min exposure are the average values from two experiments. Experiments similar to the one illustrated in Fig. 3

neuron damage sooner than SD of normoxic tissue. Moreover, cerebral tissue perfused by normally oxygenated blood tolerates protracted depolarization better than similar tissue maintained in vitro and supplied by well-oxygenated ACSF. To find what it is in blood that protects neurons against protracted depolarization is a challenge for the future. The hypothetical protective factor is absent in ACSF, and its supply is interrupted in ischemic tissue.

References

Balestrino M, Somjen GG (1986) Chlorpromazine protects brain tissue in hypoxia by delaying spreading depression-mediated calcium influx. Brain Res 385: 216-226

Balestrino M, Aitken PG, Somjen GG (1989) Spreading depression-like hypoxic depolarization in CA1 and fascia dentata of hippocampal slices: relationship to selective vulnerability. Brain Res 497: 102-107

Bureš J, Burešová O (1957) Die anoxische Terminaldepolarisation als Indicator der Vulnerabilität der Grosshirnrinde bei Anoxie und Ischämie. Pflüger's Arch 264: 325-334

Burešová O, Bureš J (1969) The effect of prolonged cortical spreading depression on learning and memory in rats. J Neurobiol 1: 135-146

Hansen AJ (1985) Effect of anoxia on ion distribution in the brain. Physiol Rev 65: 101-148

Kawasaki K, Czéh G, Somjen GG (1988) Prolonged exposure to high potassium concentration results in irreversible loss of synaptic transmission in hippocampal tissue slices. Brain Res 457: 322-329

Lauritzen M (1987) Cerebral blood flow in migraine and cortical spreading depression. Acta Neurol Scand 76 [Suppl 113]: 1-40

Marshall WH (1959) Spreading cortical depression of Leão. Physiol Rev 39: 239-279

Nedergaard M, Hansen AJ (1988) Spreading depression is not associated with neuronal injury in the normal brain. Brain Res 449: 395-398

Rusčáková D (1964) Effect of potassium ions on the morphological picture of cells in the cerebral cortex. Physiol Bohemoslov 13: 161-166

Siesjö BK (1981) Cell damage in the brain: a speculative synthesis. J Cereb Blood Flow Metab 1: 155-185

Somjen GG (1988) Basic mechanisms in cerebral hypoxia and stroke: background, review and conclusions. In: Somjen, G. G. (ed) Mechanisms of cerebral hypoxia and stroke. Plenum, New York, pp 447-466

Somjen GG (1991) Effects of hypoxia on neuronal membranes. In: Ohnishi S. T. (ed) Membrane linked diseases, vol 1. Taylor and Francis, New York (in press)

Somjen GG, Aitken PG (1984) The ionic and metabolic responses associated with neuronal depression of Leão's type in cerebral cortex and hippocampal formation. An Acad Bras Ciênc 56: 495-504

Somjen GG, Aitken PG, Balestrino M, Herreras O, Kawasaki K (1990) Spreading depression-like depolarization and selective vulnerability of neurons: a brief review. Stroke 21 [Suppl III]: III-179-III-183

Synergetics of Spreading Depression: Reentry Waves and Reverberators in the Rat Brain

J. Bureš[1], V. I. Koroleva[2], and L. V. Vinogradova[2]

[1]Institute of Physiology, Czechoslovak Academy of Sciences, Videnska 1083, 14220 Prague 4-Krc, Czechoslovakia
[2]Institute of Higher Nervous Activity and Neurophysiology, Academy of Sciences, Moscow, USSR

Introduction

Discovery of spreading depression (SD; Leão 1944) demonstrated not only a new mechanism of nonsynaptic interaction of neurons mediated by changes of brain microenvironment (Grafstein 1956; Marshall 1959; Bureš et al. 1974; Nicholson and Kraig 1981; Martins-Ferreira 1983) but also a new form of activity of vast neuronal populations governed by simple rules describing propagation of excitation in networks of connected excitable elements (Wiener and Rosenblueth 1946). The theory formulated by these authors leads in the case of SD to the following postulates: (1) An SD wave once initiated spreads with constant amplitude and velocity in all directions. (2) Any point of the excitable tissue is always in one of three states: (a) resting, (b) depressed, or (c) refractory. (3) The boundary between resting and depressed tissue corresponds to the wave front, the boundary between refractory and resting tissue to the wave rear. (4) The distance between wave front and wave rear determines the wave length of the SD process.

The above rules are essentially similar to those governing similar phenomena in other excitable media, e.g., in the malonic acid reagent (Zhabotinsky 1974), rabbit heart (Allesie et al. 1973), or aggregates of social amebae (Gerisch 1965). They account not only for the propagation of an expanding ring of excitation which expires at the boundaries of the excitable tissue, but also for more complex phenomena, the prototype of which is the rotating wave (see Winfree 1980 for review). The possibility of inducing SD rotation around an obstacle surrounded by excitable tissue has been demonstrated in the cerebral cortex of rats by Shibata and Bureš (1972, 1974a,b) and in the isolated chicken retina by Martins-Ferreira et al. (1974). The possibility of inducing SD rotation in the absence of an obstacle was rejected by Wiener and Rosenblueth (1946) but inferred by Selfridge (1948), Balakhovsky (1965), and Reshodko and Bureš (1975) on the basis of analytical arguments and computer simulations. Gorelova and Bureš (1983) produced spiral SD waves in isolated chicken retina by one of two procedures: (1) by blocking a length of the perimeter of an advancing SD wave with a DC source anode or (2) by eliciting a new SD wave in the rear of an advancing SD wave. In the first case, after the block has been removed, the two ends of the open SD wave front start spreading not only forward but also into the previously blocked area and form two incurving horns which reconstitute the close perimeter of the wave but at the same time produce (when the gap is wide enough) another open-ended segment of the wave front moving in the opposite direction and maintaining continued SD generation. In the second case, the SD-eliciting stimulus produces an open-ended segment of the SD wave front propagating into the recovering regions of retina. The free ends gradually turn around and again produce horn-shaped waves which become the nuclei of new SD generators.

Reentry Waves of SD in the Cerebral Cortex

The simplest demonstration of SD reentry in the two-dimensional sheet of rat neocortex requires that a single SD wave elicited from a definite cortical area propagates through a set of at least two electrodes first in a centrifugal and then in the opposite direction. Such "boomerang" waves of SD indicate that the regular closed wave front has been opened at some peripheral locus and that the free end(s) of the wave has turned around and penetrated through the formerly blocked area into the recovered cortex behind the advancing wave and continued to spread toward the original SD focus. Reversible SD block can be achieved by local treatment of cerebral cortex with bivalent cations (Buresová and Bureš 1960; Shibata and Bureš 1974b) but chemical block cannot be terminated fast enough to provide the reentry path for the returning SD wave. A rapidly reversible SD block can be produced by low-frequency (6-10 Hz) stimulation of cerebral cortex by brief electrical pulses (50-100 µs). Such focal stimulation elicits in the cerebral cortex of rats anesthetized with pentobarbital a sustained increase of $[K^+]_e$ to 8-10 mmol/l (Lux and Neher 1973). An increase of $[K^+]_e$ to higher levels is prevented by increased activity of the metabolic pump returning the K^+ ions released from the depolarized neurons back into the intracellular compartment. The increased K^+ clearance, demonstrated by faster removal of extrinsic potassium ions iontophoretically applied into the stimulated cortex (Heinemann and Lux 1977), accounts for the SD blockade: the potassium ions penetrating into the stimulated region from the advancing SD wave front are removed so effectively that the autoregenerative process of SD is disrupted. The block disappears during 1-3 min after cessation of stimulation (Reddy and Bureš 1980). Using this method, Koroleva and Bureš (1979) produced the stimulation block in the frontoparietal cortex and elicited SD from the occipital cortex (Fig. 1). The SD wave propagated through the temporal cortex around the block. The stimulation was discontinued shortly after the SD wave had started to penetrate into the frontal cortex. The open end of the SD wave front in the region of the disappearing block became a starting point for a new SD wave propagating not only toward the frontal pole but also into the formerly blocked area. This wave could not penetrate into the still refractory temporal cortex but propagated reliably into the already recovered parietal and occipital cortex. The open temporal end of the wave front finally approached the rhinal sulcus and disappeared before it could turn around and start a reverberating spiral wave.

The "boomerang" wave could be transformed into a reverberating or spiral wave, however, by gradually decreasing the intensity of the blocking stimulation after the SD wave had entered the frontal cortex (Fig. 2). The diameter of the block was decreased and a corridor

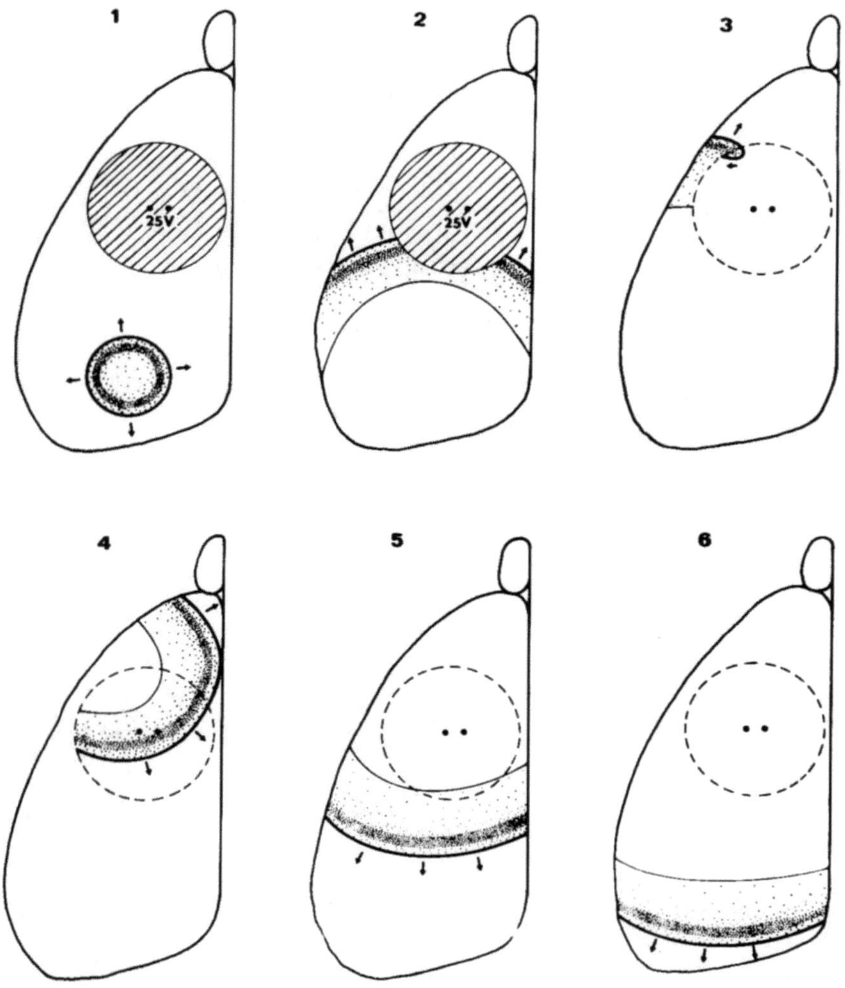

Fig. 1. Producing a reentry wave of cortical SD. Left hemisphere of the rat brain. *Dotted area* represents the SD wave, with the *thick and thin lines* denoting the wave front and waverear, respectively. The *maximum density of dots* corresponds to maximum negativity of the slow potential. *Arrows* point in the direction of SD propagation. *Shaded circular area* represents the stimulation (25 V) induced SD blockade. *S*: stimulating electrodes. The interval between individual frames is 1 min. Note that the blocking stimulation was discontinued shortly after frame 2

of SD-susceptible cortex appeared between the block and the sagittal sulcus. The SD wave propagating around the reduced block into the frontal cortex had eventually entered this corridor and spread through it into the already recovered parietal and occipital cortex but also into the temporal cortex and thus started to reverberate around the residual block. It is conceivable that after complete cessation of stimulation SD continues to reverberate in the form of a spiral wave.

Reverberation probably accounts for long series of SD waves occasionally observed after application of filter papers soaked with hypertonic KCl on the exposed cerebral cortex

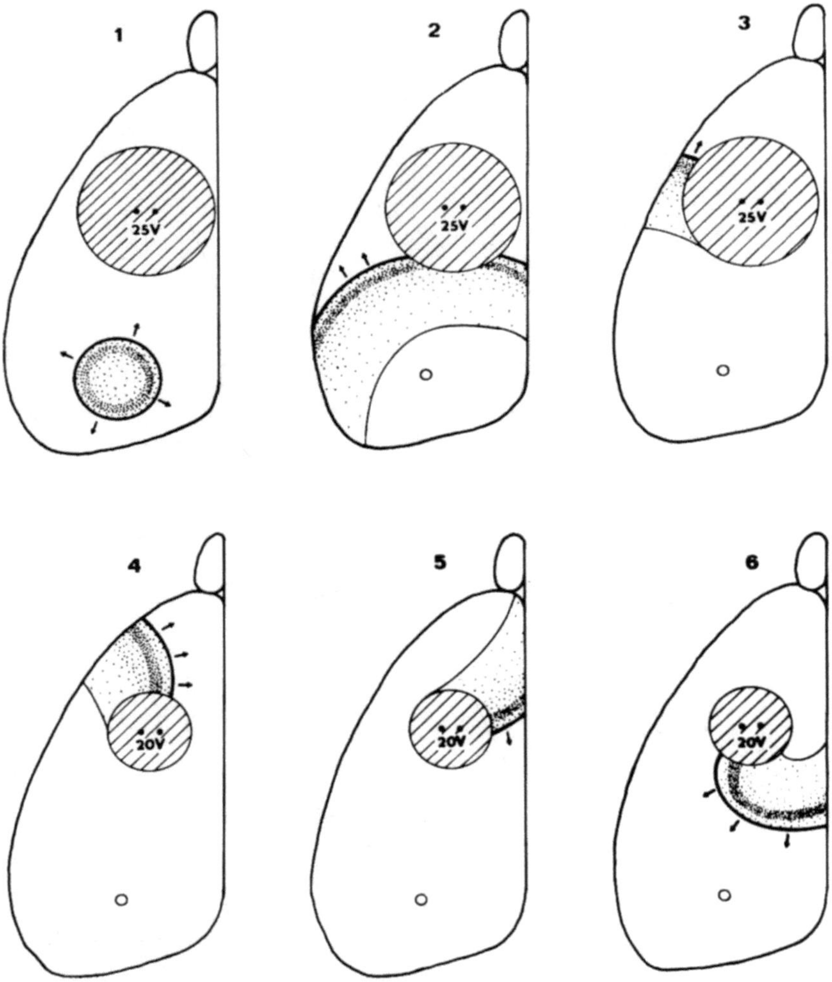

Fig. 2. Spreading depression reverberation around a stimulation-induced block. Same description as in Fig. 1. Note that the stimulation intensity was reduced from 25 to 20 V shortly after frame 3

(Bureš et al. 1974; Kuznetsova and Koroleva 1978). The depolarized cortex represents an SD focus but at the same time also an obstacle to SD propagation. Whereas the first SD wave generated immediately after KCl application starts from the entire focus the second and subsequent waves appear at discrete areas of its circumference where the surrounding cortex has had the shortest refractory period. In the example shown in Fig. 3, an SD wave appearing at the medial border of a frontoparietal KCl focus cannot spread in the rostral direction through the still refractory tissue in the narrow corridor between the focus and the sagittal sulcus, but propagates in the occipito-temporal direction around the focus. When it reaches the recovered parasagittal corridor before a new spontaneous SD waves is triggered, reverberation starts and may completely replace further generation of spontaneous SD waves. With decay of KCl concentration in the focus the blocked area gradually shrinks and the wave may turn into a spiral.

Gorelova (1986) reported that series of SD waves observed in the cerebral cortex of rats after repeated injections of potassium chloride (n = 148) or during prolonged low-frequency electrical stimulation (n = 140) are due in more than 50% cases to generation of spiral waves. Such waves appeared at a set of cortical electrodes in a characteristic sequence which could not be generated by a concentric wave propagating from a single point. The average cycle duration was 274 s. About 30% of these reverberators produced three or more cycles (maximum 18). The relatively short life of the reverberators was due to gradual

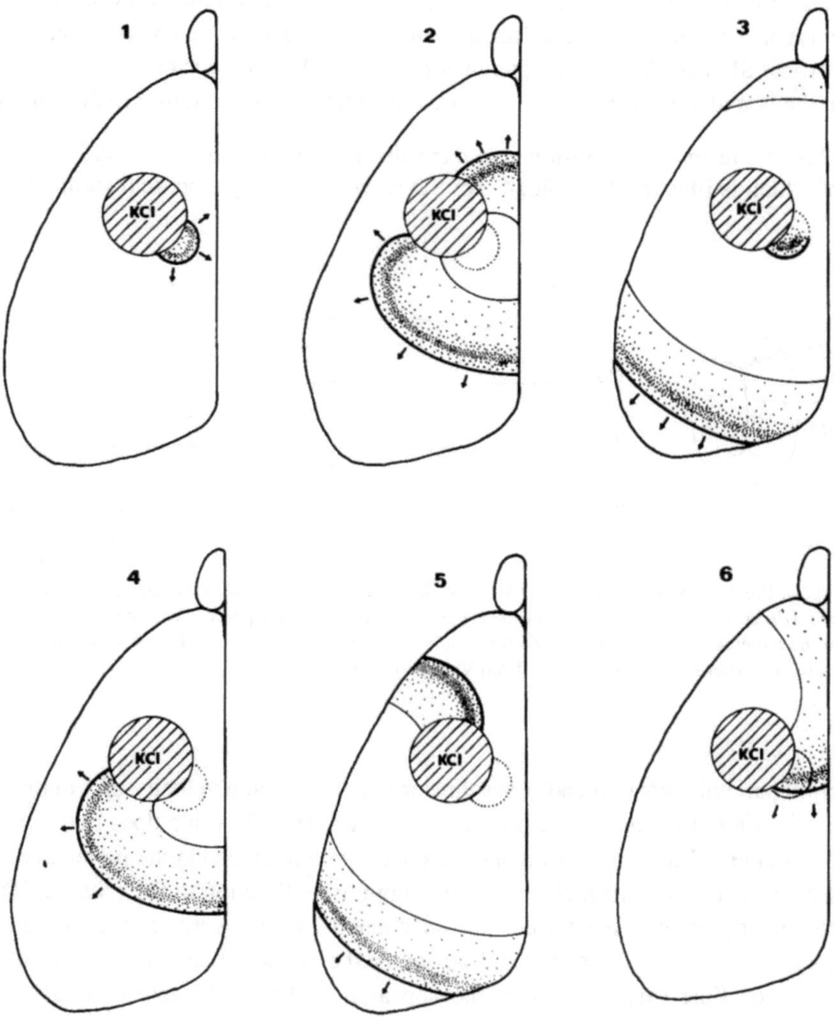

Fig. 3. Spreading depression reverberation around a nonhomogeneous KCl focus. Other description as in Fig. 1. In frame 1 a concentric SD wave is generated from the mediocaudal part of the focus. In frame 3 an open-ended SD wave propagates from the same point in the caudal direction and starts reverberation around the focus

migration of the nucleus of the spiral. Reverberation stopped when the nucleus approached the boundary of the cortex to a distance preventing SD propagation.

SD Reentry Between the Cortex and Caudate Nucleus

Whereas the reentry waves and spiral waves described in the previous section were triggered by opening the SD wave front, the same phenomena can also be generated in anisotropic excitable media supporting one-way conduction of the wave through bridges connecting large areas of homogeneous tissue. This can be illustrated in the example of SD propagation between the cerebral cortex and caudate nucleus in rats. As shown by Fifková and Syka (1964) and DeLuca et al. (1975), the two structures are connected through the amygdala by an almost continuous band of gray matter which can sustain propagation of cortical SD into the caudate-putamen complex. Most SD waves entering the caudate expire at the boundaries of the latter structure but some succeed to return to the cortex.

This was recently demonstrated in experiments (Vinogradova et al. 1991), a typical example of which is illustrated by Fig. 4. An SD wave elicited by electrical stimulation from the oc-

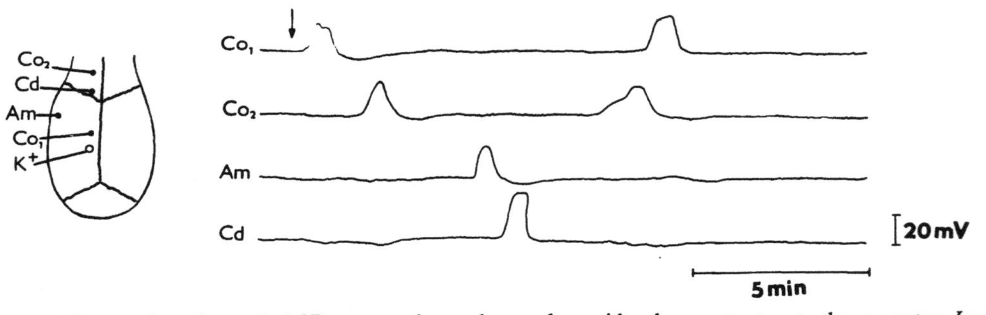

Fig. 4. Propagation of a cortical SD wave to the caudate nucleus with subsequent return to the neocortex. *Inset brain diagram* shows the position of the electrodes in the left hemisphere. *Co*$_1$, Co$_2$, Cd, Am, recording electrodes in the parietooccipital and frontal cortex, caudate nucleus, and amygdala. K^+, intracortical injection of potassium chloride. Calibration, 20 mV, 5 min (from Vinogradova et al. 1991)

cipitoparietal cortex spreads over the cortical surface at a velocity of 4 mm/min and appears 5 min after stimulation in the caudate nucleus. The SD wave does not expire here but appears after 2.5 min in the frontal cortex and 1 min later also in the parietal cortex, i.e., it passes between the cortical electrodes in an opposite direction to the original SD wave. This is convincing proof of reentry because such late SD waves were never observed in the cerebral cortex when the cortical SD did not enter the caudate nucleus. The mean corticocaudate transition time (measured from the appearance of the SD slow potential at the electrode closest to the parieto-occipital stimulation point and the caudate electrode) was significantly longer (5.9 ± 0.1 min) than the mean caudate-cortical transition time (4.8 ± 0.1 min) measured between the same electrodes. The assumption seems to be justified, therefore, that the cortical SD wave enters the caudate nucleus at one point and passes from it back to the neocortex at another point. Closer analysis of the slow potential recordings indicated that the corticocaudate transition proceeds through the amygdala whereas the caudate-cortical transition takes place at a more rostral point in the area of the claustrum.

This assumption was supported by experiments in which an SD wave was elicited by microinjection of KCl into the caudate nucleus and its propagation to the cortex was monitored by a pair of cortical electrodes placed 3 mm from the midline and 4 mm rostral and 4 mm caudal from the bregma. From the time interval between the appearance of the slow potential waves at these electrodes it is possible to determine the probable region of the caudate-cortical transition. With an SD propagation rate of 3 mm/min, a wave front normal to the line connecting the two electrodes requires 2.7 min to cover the interelectrode distance. On the other hand, an SD wave spreading from a point equidistant from both electrodes reaches them simultaneously. Since the SD wave reached the rostral electrode 1.7 min earlier than the caudal one, the caudate-cortical transition zone must be at a point 5 mm more distant from the caudal than from the rostral electrode. Such points are connected by a line corresponding to cross-sections of two circles with the radii x and $(x+5)$ with centers at the rostral and caudal electrodes, respectively. This line (Fig. 5) passes from the point *AP* –2.5, *L* 3.0 in a rostrotemporal direction and reaches the temporal boundary of the neocortex in the coronal plane – 4.0. Curvature of the frontal cortex increases the surface velocity of SD propagation in the lateromedial direction, however, and thus shifts the transition points caudally (heavy line), i.e., to the coronal plane *AP* –2.5. The most probable transition point in this region is the rostral claustrum, the more caudal parts of which are contiguous with the caudate-putamen complex.

The rostral transition point can also provide a path for propagation of cortical SD waves into the caudate but the reliability of transition through this route is much lower than transition through the amygdaloid complex. Only one-third of cortical SD waves entering the caudate nucleus used this route as indicated by the shorter corticocaudate latency (4.7 ± 0.2 min) and by the absence of reentry waves. It is only possible to speculate about the nature of this anisotropy of SD conduction at the rostral transition point. Cortical SD is accompanied by maximum $[K^+]_e$ increase in the upper cortical layers and is considerably weaker in layers 5 and 6. It is conceivable that this relatively weak stimulus cannot induce SD in the compact mass of claustrum adjacent to deep layers of neocortex. On the contrary, a high potassium increase accompanying SD in the caudate and claustrum may penetrate across the barrier between the claustrum and caudate more reliably. The anisotropy provides a reentry path for cortical SD waves which have invaded the caudate through the amygdalar route and can return to the cortex through the claustrum. The reentry does not generate reverberation in this case, however, because the cycle would be too short (about 3 min) and the circuit would include at least two loci with a low transition reliability further decreased by refractoriness. The location of the transition site in the recovered frontotemporal cortex leads to the formation of a regular concentric wave which does not initiate reverberation of SD in the neocortex.

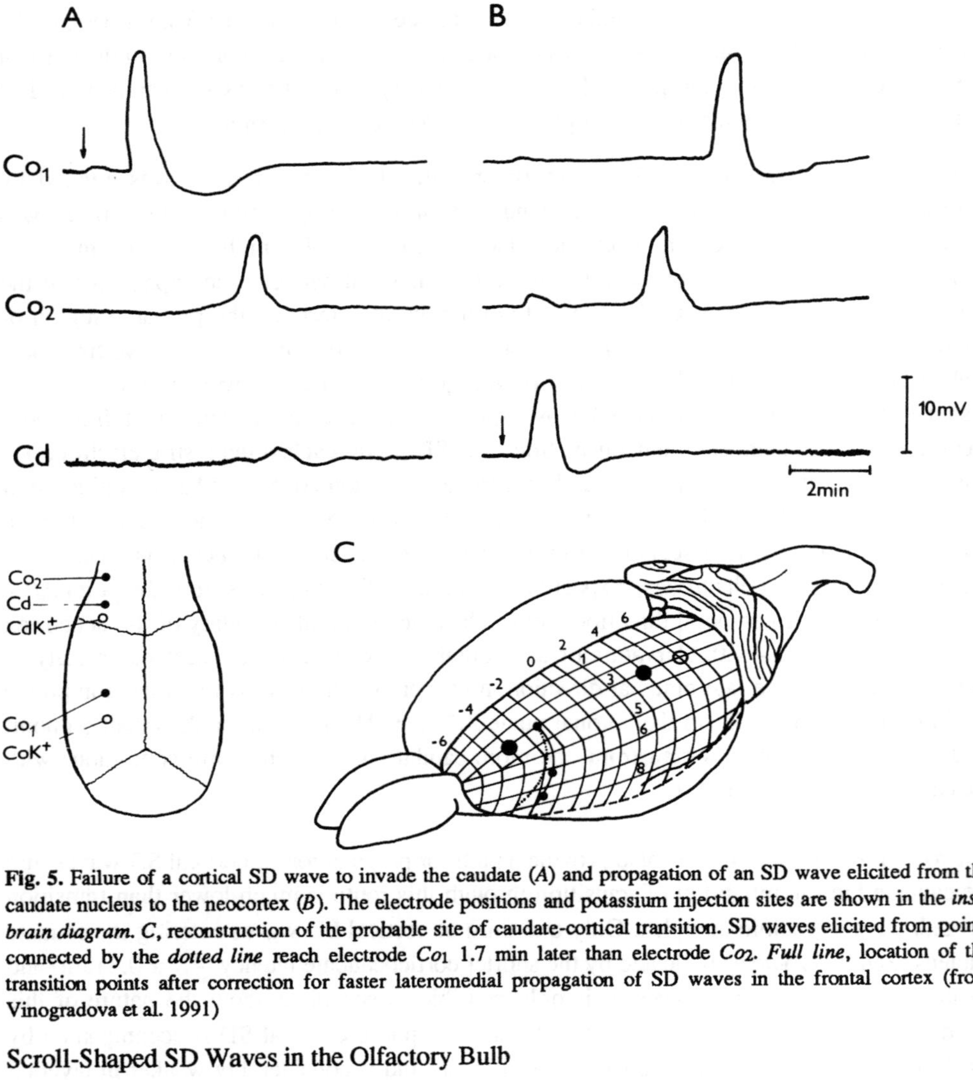

Fig. 5. Failure of a cortical SD wave to invade the caudate (*A*) and propagation of an SD wave elicited from the caudate nucleus to the neocortex (*B*). The electrode positions and potassium injection sites are shown in the *inset brain diagram*. *C*, reconstruction of the probable site of caudate-cortical transition. SD waves elicited from points connected by the *dotted line* reach electrode Co_1 1.7 min later than electrode Co_2. *Full line*, location of the transition points after correction for faster lateromedial propagation of SD waves in the frontal cortex (from Vinogradova et al. 1991)

Scroll-Shaped SD Waves in the Olfactory Bulb

Spiral waves of SD in the isolated chicken retina (Gorelova and Bureš 1983) or in the cerebral cortex of rats (Petviashvili et al. 1983) rotate around a point in a sheet of excitable tissue the thickness of which does not exceed the SD wavelength. Other SD-prone structures of mammalian and avian brain may support three-dimensional propagation of concentric spherical waves which can be changed into scroll-shaped rotating waves. In such cases the wave rotates not around a point but around a filament (Winfree 1980), which can be either open (cylindrical scroll) or closed (toroidal scroll).

The cylindrical scroll waves of SD were demonstrated for the first time in pigeon striatum (Shibata and Bureš 1980), which allows SD propagation not only over the surface but also to a depth of 5-6 mm. A vertical cylindrical lesion penetrating through the hyperstriatum and neostriatum to the paleostriatum, archistriatum, and brain stem was made by thermocoagulation in one hemisphere of the pigeon brain. The lesion was about 3 mm in diameter and was surrounded on all sides by at least 2 mm of intact striatal tissue. Reverberation was elicited by the two-injection technique similar to that used in the case of reverberating cortical SD (Shibata and Bureš 1972, 1974a,b) but both injections were applied 2 mm below the brain surface (Fig. 6). The open-ended hemisheric wave produced by the second injection reached the upper and lower boundaries of the tissue before it could turn

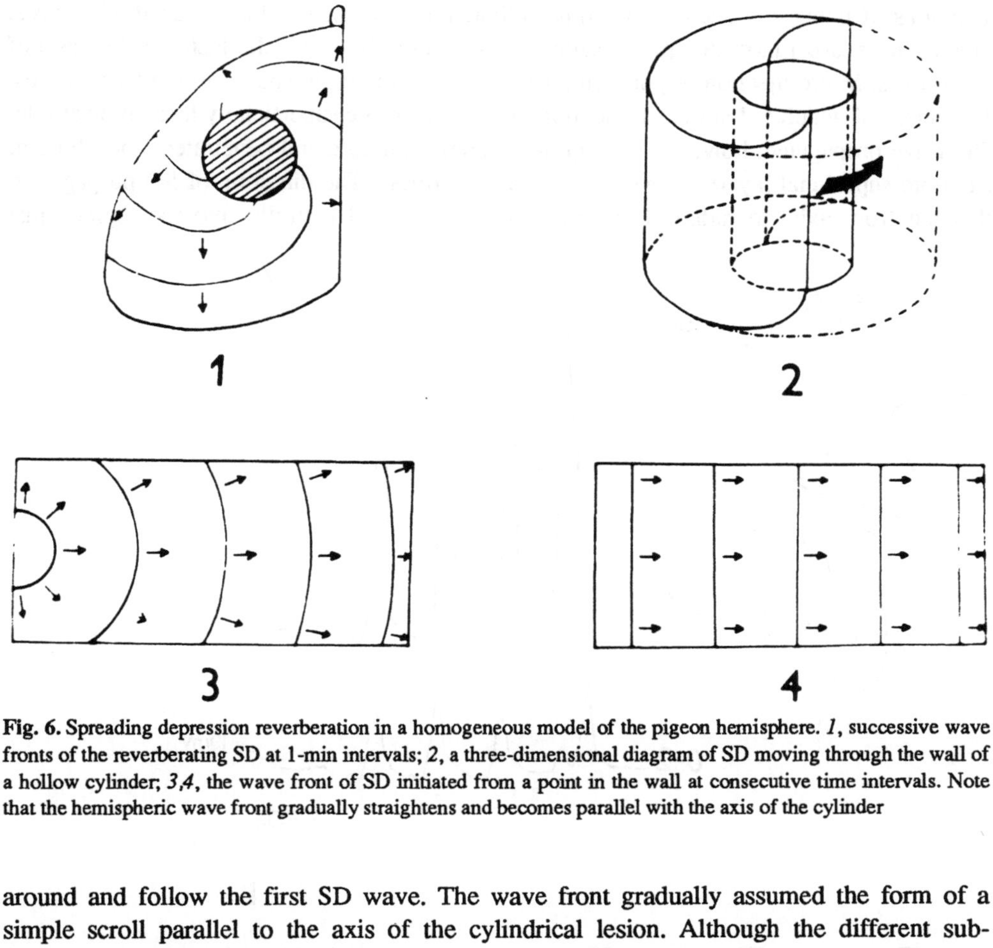

Fig. 6. Spreading depression reverberation in a homogeneous model of the pigeon hemisphere. *1*, successive wave fronts of the reverberating SD at 1-min intervals; *2*, a three-dimensional diagram of SD moving through the wall of a hollow cylinder; *3,4*, the wave front of SD initiated from a point in the wall at consecutive time intervals. Note that the hemispheric wave front gradually straightens and becomes parallel with the axis of the cylinder

around and follow the first SD wave. The wave front gradually assumed the form of a simple scroll parallel to the axis of the cylindrical lesion. Although the different subdivisions of avian striatum are separated by layers of myelinated fibers, decreasing the reliability of SD propagation, reverberation stopped after 6.3 ± 1.6 cycles (n = 12) on average. No attempt was made to elicit scrolls in the intact pigeon brain, but existence of such waves has been recently demonstrated in the olfactory bulb of rats.

The olfactory bulb belongs to neural structures with low SD susceptibility which must be increased by superfusion of the exposed brain surface with sodium acetate (Leão 1963; Leão and Martins-Ferreira 1963). Amemori et al. (1987) showed that bulbar SD in rats can be reliably elicited by microinjection (0.5-1.0 μ l) of isotonic potassium acetate into the granule cell layer. The slow potential wave, which attained an amplitude of 25 mV, propagated with a velocity of 3-4 mm/min through all layers of the olfactory bulb. The amplitude decreased in the rostrocaudal direction and the wave expired at the caudal boundary of the olfactory bulb.

SD waves elicited from the rostral pole of the olfactory bulb advanced through all layers of this ellipsoid structure and attained maximum negativity in its granule cell core. Repeated injections of potassium acetate sometimes elicited a series of regularly spaced SD waves, suggesting reverberation. A typical example is shown in Fig. 7, which illustrates the onset of the process. Microinjection of potassium acetate elicited in rapid succession two SD waves, the second of which did not reach the more distant and more superficially located electrode. The wave propagated, however, through the granule cell core and penetrated from there to the more superficial layers recovered from refractoriness. The direction of SD propagation changed from axial to radial. The wave could not spread rostrally into the regions just

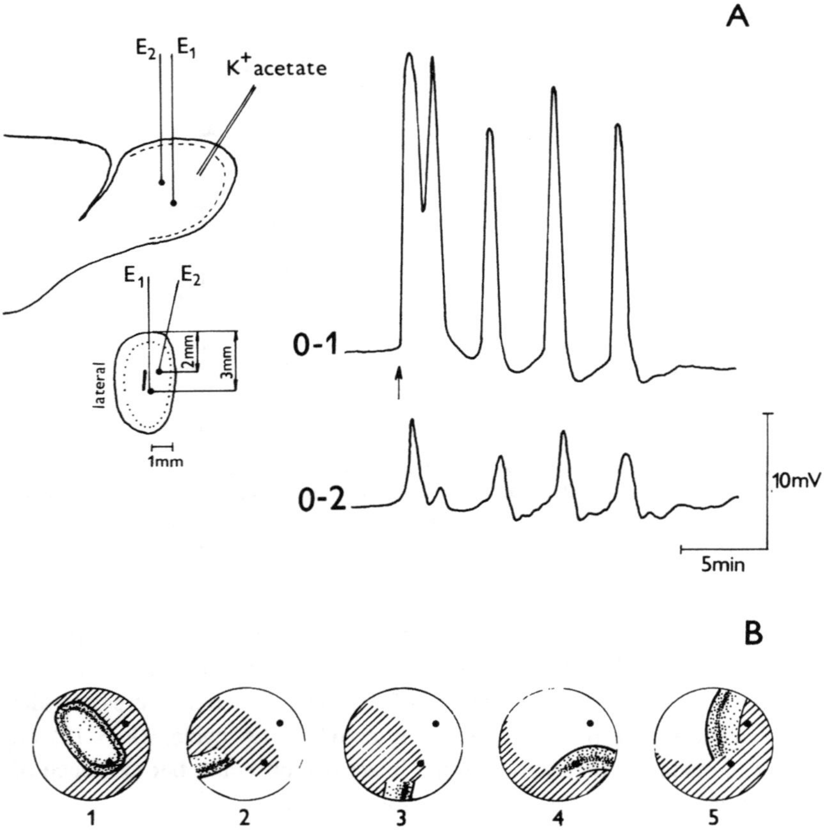

Fig. 7A,B. Example of repetitive SD waves elicited in the rat olfactory bulb by a single injection of potassium acetate. **A** Recording with two microelectrodes 1 and 2 the positions of which are shown in the *inset brain scheme*. **B** Development of the scroll wave on the frontal section of the bulb. SD refractoriness is indicated by *shading*. Other description as for Fig. 1. For details see text. (Modified from Amemori et al. 1987)

passed by the second SD wave but entered recovered parts of the olfactory bulb in the same and more rostral coronal planes. The radius started to rotate around the axis of the bulb, which was manifested by the fact that the wave returned after 4 min to the lower and 20 s later to the upper electrode. This cycle was repeated three times at regular 4 min intervals before it stopped. With the interelectrode distance approximately 1.0 mm and SD velocity 3 mm/min, such a delay indicates that the wave front was perpendicular to the line connecting the two electrodes and was clearly different from the wave front elicited by microinjection of potassium acetate, which appeared at both electrodes simultaneously.

The presence of SD reverberation in the olfactory bulb was indicated not only by the slow potential waves but also in the unit activity recordings (Amemori and Bureš 1988). Figure 8 shows repetitive unit activity bursts appearing at regular 5-min intervals in the mitral cell layer of a caudal part of the bulb as a late aftermath of a single injection of potassium acetate into the rostral pole of the olfactory bulb. Since no slow potential changes accompanied the bursts, the unit recording site was obviously affected by the remote influence of

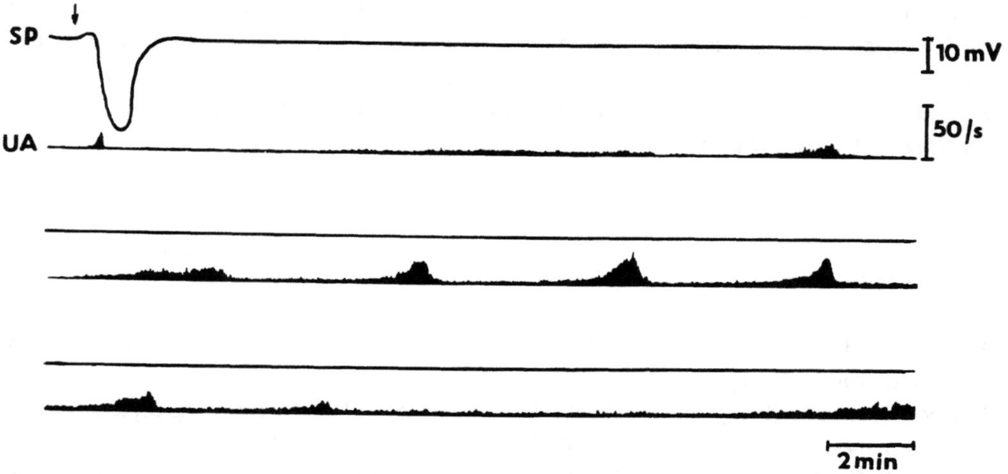

Fig. 8. Repetitive discharges of an olfactory bulb unit following a series of SD waves probably corresponding to SD reverberation in a remote part of the olfactory bulb. *UA*, unit activity integrated at 1-s intervals. *SP*, slow potential wave of SD. Calibration, 10 mV, negativity downward for the slow potential; 50/s, 50 impulses/s for the unit activity record. *Time scale*, 2 min (from Amemori and Bureš 1988)

SD waves circulating in the more rostral parts of the olfactory bulb not monitored by the capillary electrode. The 5-min duration of the cycle is consonant with the notion that this activity is generated by a scroll wave, the axis of which corresponds to the rostrocaudal axis of the bulb. With an olfactory bulb diameter of 4 mm and a circumference of 12-13 mm, a cycle duration of 5 min corresponds to a propagation velocity of 2.4-2.6 mm/min. This relatively low value reflects the fact that the scroll wave always propagates through tissue not yet fully recovered from the preceding wave. It is conceivable that the scroll wave propagates only through a rostral cone-shaped segment of the bulb and does not penetrate into the more caudal segments.

Conclusions

The fact that SD satisfies the definition of autowaves studied by synergetics makes it possible to antecipate the existence of unusual neural phenomena and to demonstrate them experimentally. Small dimensions and inhomogeneity of the excitable tissue account for the failure of the rat brain to generate the full range of reverberators (e.g., anomalous reverberators, multiarmed vortices) or to multiply reverberators at a rate leading to chaotic turbulence (Krinsky 1984). Nevertheless, the examples reviewed above show that the mammalian CNS can support elements of SD reverberation in two-dimensional (cerebral cortex) as well as in three-dimensional (cortex-caudate, olfactory bulb) structures. Such phenomena represent a purely functional mechanism changing a single-cycle reversible process into a generator of repetitive activity sustained by processes mediating the recovery of the initial high-energy level of the tissue. It is somewhat paradoxical that homeostasis of the brain microenvironment which restores the resting state after each individual SD wave helps to maintain the existence of SD once SD reverberation has been initiated. Autowave sources of this type may play a role in epilepsy or migraine and can contribute to the symptomatology of head injuries and focal brain ischemia. SD research has pioneered the view that the brain is a nonequilibrium, self-organized, synergistic system. It can be expected that the synergetic approach will improve our understanding of other physiological and pathological aspects of brain function based on inadvertent interaction within massive aggregates of active neurons.

References

Allesie MA, Bonke FIM, Schopman FJG (1973) Circus movement in rabbit atrial muscle as a mechanism of tachycardia. Circ Res 33: 54-62

Amemori T, Bureš J (1988) Functional ablation of the olfactory bulb by spreading depression: unit activity changes and transient anosmia. Brain Res Bull 20: 421-427

Amemori T, Gorelova NA, Bureš J (1987) Spreading depression in the olfactory bulb of rats: reliable initiation and boundaries of propagation. Neuroscience 22: 29-36

Balakhovsky IS (1965) Some forms of excitation spread in an ideal excitable tissue (in Russian). Biofizika 10: 1063-1067

Bureš J, Burešova O, Křivánek J (1974) The mechanism and applications of Leão's spreading depression of EEG activity. Academic, New York

Burešová O, Bureš J (1960) The use of partial functional decortication in the study of the localization of conditioned reflexes. Physiol Bohemoslov 9: 210-218

DeLuca B, Shibata M, Brozek G, Bureš J (1975) Facilitation of Leão's spreading depression by a pyrrolopyrimidine derivative. Neuropharmacology 14: 537-545

Fifková E, Syka J (1964) Relationships between cortical and striatal spreading depression in rat. Exp Neurol 9: 355-366

Gerisch G (1965) Stadienspezifische Aggregationsmuster bei *Dyctyostelium discoideum*. Wilhelm Roux Arch Entwicklungsmechanik Org 156: 127-144

Gorelova NA (1986) Spiral waves of spreading depression in the central nervous system (in Russian). PhD Thesis, Institute of Higher Nervous Activity and Neurophysiology, Academy of Sciences, Moscow

Gorelova NA, Bureš J (1983) Spiral waves of spreading depression in the isolated chicken retina. J Neurobiol 14: 353-363

Grafstein B (1956) Mechanism of spreading cortical depression. J Neurophysiol 19: 154-171

Heinemann U, Lux HD (1977) Ceiling of stimulus-induced rises of extracellular potassium concentration in the cerebral cortex of cat. Brain Res 120: 231-249

Koroleva VI, Bureš J (1979) Circulation of cortical spreading depression around electrically stimulated areas and epiletic foci in the neocortex of rats. Brain Res 173: 209-215

Krinsky VI (1984) Autowaves: results, problems, outlooks. In Krinsky VI (ed) Self-organization, autowaves and structures far from equilibrium. Proceedings of an International Symposium Pushchino, 18-23 July 1983. Springer, Berlin, Heidelberg New York, pp 9-19 (Springer series in synergetics, vol 28)

Kuznetsova GD, Koroleva VI (1978) Foci of stationary excitation in the cerebral cortex (in Russian). Nauka, Moscow

Leão AAP (1944) Spreading depression of activity in cerebral cortex. J Neurophysiol 7: 359-390

Leão AAP (1963) On the spread of spreading depression. In Brazier MAB (ed) Brain Function, vol 1. University of California Press, Berkeley, pp 78-85

Leão AAP, Martins-Ferreira H (1963) Nota acerca de variação lenta de voltagem da depressão alastrante no bulbo olfativo. An Acad Bras Ciênc 35: I-II

Lux HD, Neher E (1973) The equilibration time course of $[K^+]_e$ in cat cortex. Exp Brain Res 17: 190-205

Marshall WH (1959) Spreading cortical depression of Leão. Physiol Rev 39: 239-279

Martins-Ferreira H (1983) Spreading depression in the chicken retina. In Ookawa T (ed) The brain and behavior of the fowl. Japan Scientific Societies Press, Tokyo, pp 317-333

Martins-Ferreira H, De Oliveira Castro G, Struchiner CJ, Rodrigues PS (1974) Circling spreading depression in isolated chick retina. J Neurophysiol 33: 773-783

Nicholson C, Kraig, RP (1981) The behavior of extracellular ions during spreading depression. In Zeuthen T (ed) The application of ion-selective electrodes. Elsevier, Amsterdam, pp 217-238

Petviashvili VI, Koroleva VI, Gorelova NA, Podolets A (1983) On the possibility to elicit reverberating SD waves in the cerebral cortex. In Russinov VS (ed) Electrophysiological investigations of stationary activity in the brain. Nauka, Moscow, pp 144-157

Reddy MM, Bureš J (1980) Cortical $[K^+]_e$ and the stimulation induced blockade of spreading depression in the rat cerebral cortex. Neurosc Lett 17: 243-247

Reshodko LV, Bureš J (1975) Computer simulation of reverberating spreading depression in a network of cell automata. Biol Cyb 18: 181-189

Selfridge O (1948) Studies of flutter and fibrillation. V. Some notes on the theory of flutter. Arch Inst Cardiol Mex 18: 177-187

Shibata M, Bureš J (1972) Reverberation of cortical spreading depression along closed loop pathway in rat cortex. J Neurophysiol 35: 381-388

Shibata M, Bureš J (1974a) Optimum topographical conditions for reverberating cortical spreading depression in rats. J Neurobiol 5: 107-118

Shibata M, Bureš J (1974b) Functional decortication employing reverberating cortical spreading depression: experimental evaluation of advantages and limitations. Exp Neurol 45: 415-423

Shibata M, Bureš J (1980) Reverberating spreading depression in pigeon forebrain. In: Tanabe Y, Tanaka K, Ookawa T (eds) Biological rhythms in birds: neural and endocrine aspects. Japan Scientific Societies Press, Tokyo, pp 113-120

Vinogradova LV, Koroleva VI, Bureš J (1991) Reentry waves of Leão's spreading depression between neocortex and caudate nucleus. Brain Res 538: 161-164

Wiener N, Rosenblueth A (1946) The mathematical formulation of the problem of conduction of impulses in a network of connected excitable elements, specifically in cardiac muscle. Arch Inst Cardiol Mex 6: 205-265

Winfree AT (1980) The geometry of biological time. Springer, Berlin Heidelberg, New York

Zhabotinsky AM (1974) Spontaneously oscillating concentrations (in Russian). Nauka, Moscow

Recent Studies on Retinal Spreading Depression *

R. J. do Carmo and H. Martins-Ferreira

Departamento de Neurobiologia, Instituto de Biofísica Carlos Chagas Filho,
Universidade Federal do Rio de Janeiro, Ilha do Fundão, 21941, RJ, Brasil

Introduction

A number or pathological conditions of the brain, such as ischemia, hypoxia, and epilepsy have been shown to be associated with the accumulation of unsaturated fatty acids from the cellular membranes (a recent account of the present state of art is found in Barkai and Bazan 1989), the metabolism of which involves a cascade of chemical reactions and the formation of highly reactive free radical species. These elements cause peroxidative alteration to the membrane fatty acids leading to varying degrees of cell damage (Demopoulos 1972).

A wealth of experimental evidence, accumulated over the years, points to a transient breakdown of the selective permeability of the neural cell membrane during the phenomenon of retinal spreading depression (SD) (Martins-Ferreira and do Carmo 1987). Considering these results, together with the above-mentioned action of free radicals on cell membranes, Netto and Martins-Ferreira (1989) succeeded in producing SDs in retinal preparations by rose bengal photodynamic action without damaging the tissue.

Recently, Lauritzen et al. (1990), in a study of the energy state in the cerebral cortex invaded by SD, reported a marked increase in concentration of arachidonic acid, coincident with the phenomenon. This observation reinforces the suspicion that free radicals are fundamentally involved in the mechanism of SD.

In the present work we continued these investigations by studying the effects of unsaturated fatty acids, exogenously administered, on isolated retinas.

Methods

The experiments were performed on white leghorn chick retinas, either with or without pigment epithelium. Enucleation proceeded immediately after decapitation, and the eyeballl was sectioned along its equator. The vitreous humor was withdrawn and the posterior part of the eye was cut in strips of about 4 × 10 mm, which were then immersed in Ringer's solution. The retina of one of the strips was detached from the choroid and mounted with the vitreous side up in an experimental chamber (0.2 ml capacity) as shown in Fig. 1. The

*This study was supported by CNPq, FINEP, CEPG/UFRJ, and the Health Foundation.

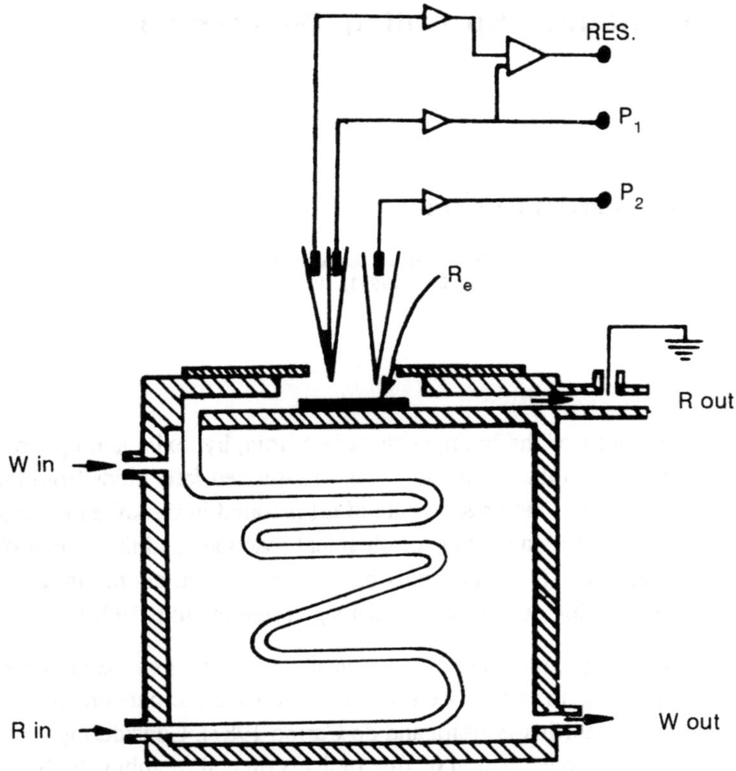

Fig. 1. Cross-section (side view) of the bathing chamber and schematic diagram of the double-barreled and single-barreled microelectrode showing the circuit arrangement for recording of slow potential (*P*) and extracellular ionic activities (*Res*)., *R*, Ringer's solution; *W*, water; *Re*, retina

preparation was superfused with a standard Ringer's solution that had the following composition (m*M*): NaCl 100, KCl 6, $CaCl_2$ 1, NaH_2PO_4 1, $NaHCO_3$ 30, $MgSO_4$ 1, and glucose 30. In some experiments Ringer's solution containing 70 m*M* NaCl and 30 m*M* sodium isethionate, the remaining constituents being kept unaltered, was used.

The solutions were equilibrated with 95% O_2/5% CO_2 and delivered to the chamber by an infusion pump at a rate of 0.6 ml/min. The temperature of the chamber was maintained constant at about 30°C.

Generally, SD was mechanically elicited every 15 min by pricking the retina with a sharpened tungsten wire in a region close to the aperture for the exit of the bathing solution.

Ringer's solutions containing a higher KCl concentration for chemical stimulation, or other substances to be tested, were administered by means of a bypass system coupled to the main superfusing circuit.

Extracellular recording of the slow potential and ionic changes occurring during SD have been described in detail elsewhere (do Carmo and Martins-Ferreira 1984). Briefly, the potential change was recorded either by means of a single-barreled microelectrode or by the potential-sensing barrel (filled with 150 m*M* NaCl) of a double-barreled ion-selective microelectrode. The extracellular activities of K^+($[K^+]_o$) and Ca^{2+} ($[Ca^{2+}]_o$) were recorded by the ion-sensing barrel containing the respective ion-exchanger (Corning 477317 for K^+ and Fluka Ca^{2+} cocktail for Ca^{2+}) at the tip and backfilled with the salt (150 m*M*) of the ion to be sensed. All potentials recorded in the proximal retina were measured against a grounded silver/silver chloride/agar/Ringer's indifferent electrode.

Ringer's solutions in which the K^+ was elevated to 10, 15, 30, and 60 m*M* or the Ca^{2+} concentration was decreased to 0.5, 0.1, and 0.05 m*M* were used as calibration solutions for K^+ and Ca^{2+} ion-selective microelectrodes, respectively.

The velocity of SD propagation was measured between two microelectrodes placed in the retina, orthogonally aligned to the visible wavefront.

For testing, the unsaturated fatty acids (0.05-0.1 ml) were added to Ca^{2+}-free, 95% O_2/5% CO_2-saturated Ringer's solution containing the detergent polyoxyethylenesorbitol monooleate (0.05%). Immediately thereafter, the resulting micelles were dispersed by sonication. Then the Ringer's solution was completed by adding $CaCl_2$ under continuous bubbling with the O_2 95%/CO_2 5% mixture. This stock suspension was slightly whitish, but devoid of signs indicative of Ca^{2+} flocculation.

At this point it should be said that the expression "concentration of fatty acids," as applied here, refers to the quantity of the substance detergent-dispersed in Ringer's solution. Since in this condition the fatty acids orientate into micelles, which may result in a change of activity, the effective concentration of the fatty acids under the present experimental conditions is not precisely known.

Control experiments have shown that Ringer's solutions containing 0.05% polyoxyethylenesorbitol monooleate do not affect SD.

Arachidonic acid (approx. 99% from porcine liver), oleic acid, linoleic acid, linolenic acid, superoxide dismutase polyethylene glycol (from bovine erythrocytes), catalase (from bovine liver), quinacrine, nordihydroguairetic acids, and polyoxyethylenesorbitol monooleate were obtained from Sigma Chemical Co., (St Louis, Mo, USA) indomethacin and dexamethasone from Merck Sharp Dohme (São Paulo, Brazil) and chlorpromazine from Rhodia (São Paulo, Brazil).

Results

Effects of Exogenous Administration of Unsaturated Fatty Acids

These experiments were undertaken to provide information related to the capacity of exogenously administered unsaturated fatty acids to initiate SD in in vitro retina preparations. In the course of the investigation it was noticed that the fatty acids reduced the activity of ionized calcium (Ca^{2+}), when added to the Ringer's solution. In view of this fact, the level of calcium chloride in the Ringer's solutions containing the fatty acids, except when otherwise required, was raised until the baseline concentration level of Ca^{2+} was attained, as indicated by means of a Ca^{2+}-selective microelectrode. Preliminary trials with oleic acid revealed that, at concentrations of about 10-50 m*M*, this fatty acid, after a variable incubation time (about 2-10 min) starts SD, but, in addition, produces several whitish patches in the preparations, indicating irreversible damage in the tissue. Even though the affected regions were not invaded by subsequently stimulated SDs, every episode of the reaction further expanded the lesions which, in many cases, took up the whole fragment of retina. At lower concentrations (1-1.6 m*M*), however, in most of the preparations it was frequently possible, with a single dose, to trigger one or several sucessive episodes of SD without damaging the tissue.

Similar results were obtained with linoleic, linolenic, and arachidonic acids (Figs. 3A, 5, 6). Arachidonic acid was more potent, being, in some experiments, effective at a concentration of about 300 μ*M*. As a rule, in the absence of the unsaturated acids, the mechanical stimulus was only able to elicit a single SD wave. However, when applied in the presence of the acids, such a mechanical stimulus usually produced a sequence of two or more episodes of SD (Figs. 2, 3B, 4B).

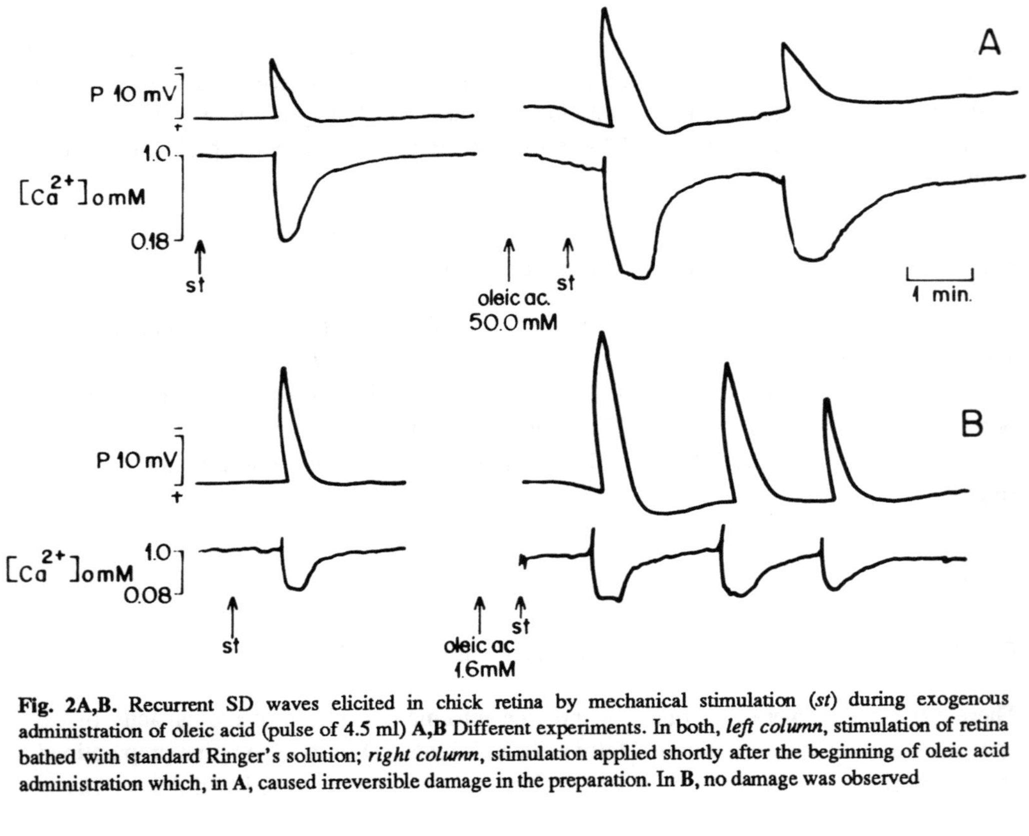

Fig. 2A,B. Recurrent SD waves elicited in chick retina by mechanical stimulation (*st*) during exogenous administration of oleic acid (pulse of 4.5 ml) **A,B** Different experiments. In both, *left column*, stimulation of retina bathed with standard Ringer's solution; *right column*, stimulation applied shortly after the beginning of oleic acid administration which, in **A**, caused irreversible damage in the preparation. In **B**, no damage was observed

It was observed that at a Ca^{2+} concentration lower than the baseline level, the effects of the fatty acids were more intense (Figs. 4, 5).

Usually these substances, in addition to lowering the amplitude of the slow potential and Ca^{2+} and K^{+} activity changes measured during SD, decreased the duration of these parameters. This was true when the acids were administered at concentrations insufficient to

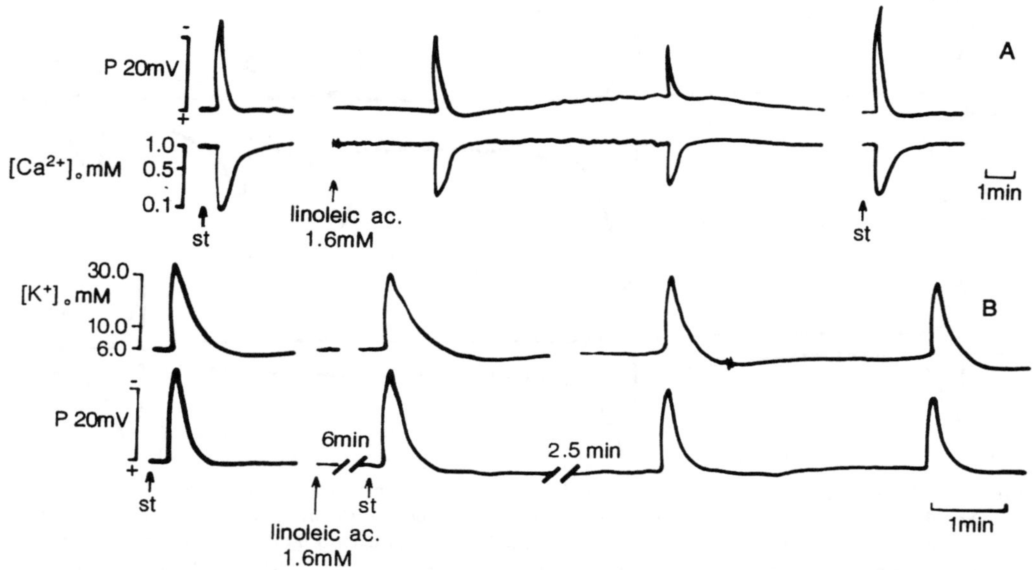

Fig. 3A,B. Exogenous administration (pulse of 4.5 ml) of linoleic acid in chick retina. **A** *Middle section*, recurrent SD waves elicited by linoleic acid; *left and right sections*, single episodes of SD elicited by mechanical stimulation (*st*) of the retina bathed with standard Ringer's solution. **B** *Left section*, as in A; *right section*, recurrent SD waves elicited by mechanical stimulation (*st*) of the retina bathed with Ringer's solution containing linoleic acid

damage the preparation (Figs. 2B, 3A,B, 4B, 6). Otherwise the potentials displayed a much slower time course and a lower amplitude (Fig. 2A). In fact, these last changes, when not very much pronounced, provided the first evidence that the preparation was deteriorating. It should be mentioned, however, that sometimes, in experiments where the mechanical stimulation was timed to coincide with the beginning of the administration of the fatty acids, both the slow potential and the ionic changes transiently increased in amplitude and duration (Fig. 2A,B).

As regards the velocity of propagation, our study was restricted to oleic and linolenic acids. Since changes in Ca^{2+} alter the speed of SD (do Carmo and Martins-Ferreira 1988), for these experiments the baseline Ca^{2+} concentration level in the solutions containing the fatty acids was maintained under strict control.

When the mechanical stimulus to initiate the phenomenon was properly timed, i.e., applied within about 1-2 min of treatment, its velocity of propagation displayed a biphasic (incre-

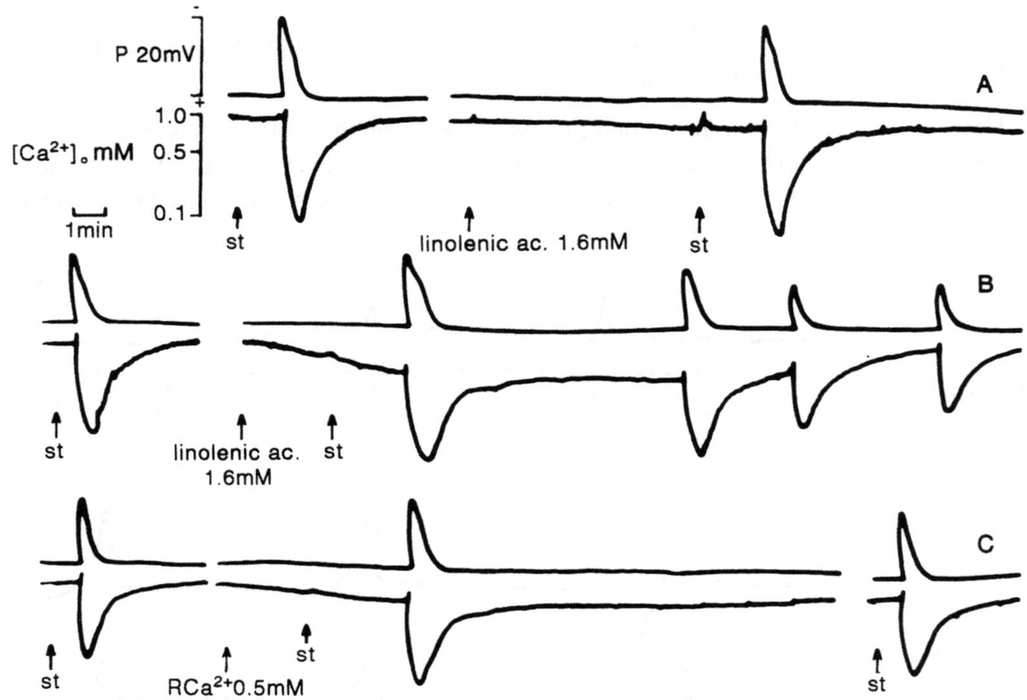

Fig. 4A-C. Sections taken from a continuous recording of one experiment in chick retina showing the effect of mechanical stimulation (*st*) in three different experimental conditions. A Before (*left*) and during (*right*) the administration of Ringer's solution containing linolenic acid [$Ca^{2+}{}_o$] adjusted to 1.0 m*M*). B Same as in A except that [Ca^{2+}] was not adjusted for decrease caused by the presence of linolenic acid. C Before (*left*), during (*middle*), and after (*right*) administration of low [Ca^{2+}] Ringer's solution

ment/decrement) change as illustrated in Fig. 7. If the stimulation was performed later, only a reversible decrease of speed was observed.

Effect of Agents that Interfere with the Liberation or the Metabolism of Unsaturated Fatty Acids

The SD generation capacity of the unsaturated fatty acids described in the preceding section led us to undertake experiments to determine the effects on SD of agents that either impede their liberation from the tissue or interfere with their metabolism (Tappel et al. 1953; Flower 1974, 1988; Hamberg 1976; Blackwell et al. 1977; Hirata et al. 1980; Lapetina et al. 1981; Billah et al. 1982; Chan et al. 1982; Bindoli et al. 1988). The following substances were used for these experiments: quinacrine (10-80 μ*M*), indomethacin (60-200 μ*M*), chlorpromazine (520 μ*M*-1.6 m*M*), dexamethasone (180-200 μ*M*), and nordihydroguaiaretic acid (230 μ*M*-1.6 m*M*).

Here again it was observed that, with the exception of quinacrine and indomethacin, the substances under consideration reduced the free Ca^{2+} concentration level in the Ringer's

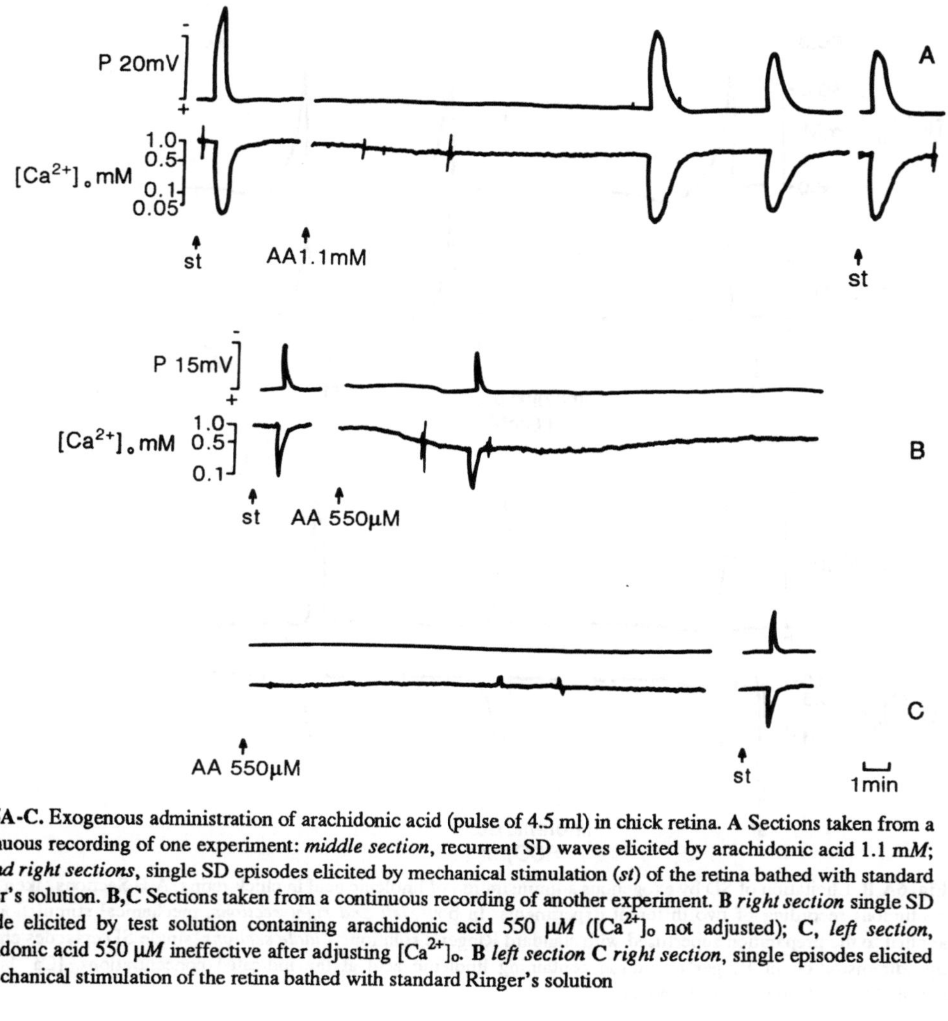

Fig. 5A-C. Exogenous administration of arachidonic acid (pulse of 4.5 ml) in chick retina. **A** Sections taken from a continuous recording of one experiment: *middle section*, recurrent SD waves elicited by arachidonic acid 1.1 m*M*; *left and right sections*, single SD episodes elicited by mechanical stimulation (*st*) of the retina bathed with standard Ringer's solution. **B,C** Sections taken from a continuous recording of another experiment. **B** *right section* single SD episode elicited by test solution containing arachidonic acid 550 μ*M* ($[Ca^{2+}]_o$ not adjusted); **C**, *left section*, arachidonic acid 550 μ*M* ineffective after adjusting $[Ca^{2+}]_o$. **B** *left section* **C** *right section*, single episodes elicited by mechanical stimulation of the retina bathed with standard Ringer's solution

solution. Therefore, the trials were performed only after the free Ca^{2+} in the test solutions was restored to the baseline level.

A preliminary exploration was performed with chlorpromazine, nordihydroguaiaretic acid, and indomethacin, to investigate their influence on SD initiation by different kinds of stimuli. The tests revealed that these agents attenuate the capacity of optical, mechanical, and chemical K^+ stimuli to elicit SD. The effect of chlorpromazine on optical and chemical stimulation is illustrated in Fig. 8.

In relation to their influence on the velocity of propagation of the reaction, all substances, in general, reversibly decreased it. Quinacrine, however, usually displayed a diphasic effect, i.e., an initial increase followed by a more pronounced decrease. As to nordihydroguaiaretic acid, the decrease in velocity lasted longer and the reversibility was partial. These effects are illustrated in Fig. 9.

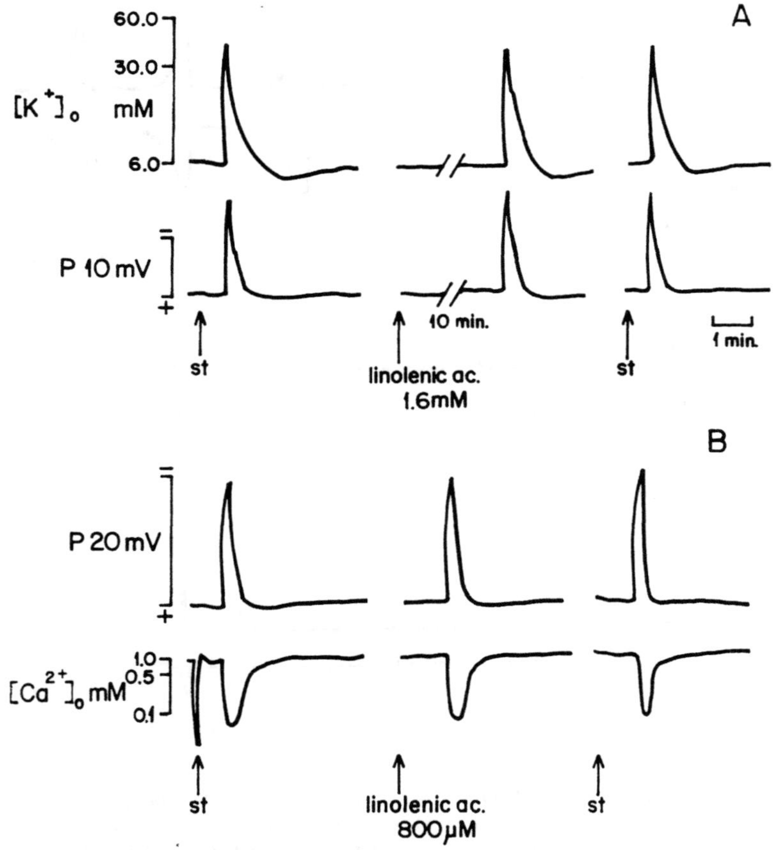

Fig. 6A,B. Elicitation of SD by exogenous administration of linolenic acid in chick retina. **A,B** Sections taken from continuous recording of two different experiments. In both, *left and right sections*, mechanical stimulation (st) applied to the preparation superfused with standard Ringer's solution. *Middle sections*, single SD episodes elicited by administration of Ringer's solution containing linolenic acid at the indicated concentrations. $[Ca^{2+}]_o$ was adjusted to 1 m*M* in both experiments

The potential and ionic $[K^+]_o$ and $[Ca^{2+}]_o$ changes in SD were also affected by the substances. They decreased the amplitude and duration of these parameters. Fig. 10 illustrates these effects on $[Ca^{2+}]_o$ and on slow potential changes.

Finally, we have tested superoxidismutase (SOD) given at an activity of 300-600 units, and catalase (100-200 m*M*) on SD. For these experiments it was found convenient to employ a small chamber as described by Vercesi and Martins-Ferreira (1983). In this closed system, tests made with insaturated fatty acids (only linolenic and arachidonic acids were tried) produced effects essentially similar to the ones described for the same agents when applied in preparations maintained under the ordinary superfusing conditions.

The antagonistic effect of SOD became manifest not only with the elevation of the threshold for mechanical stimulation but also by the diminution of the velocity of propagation of SD.

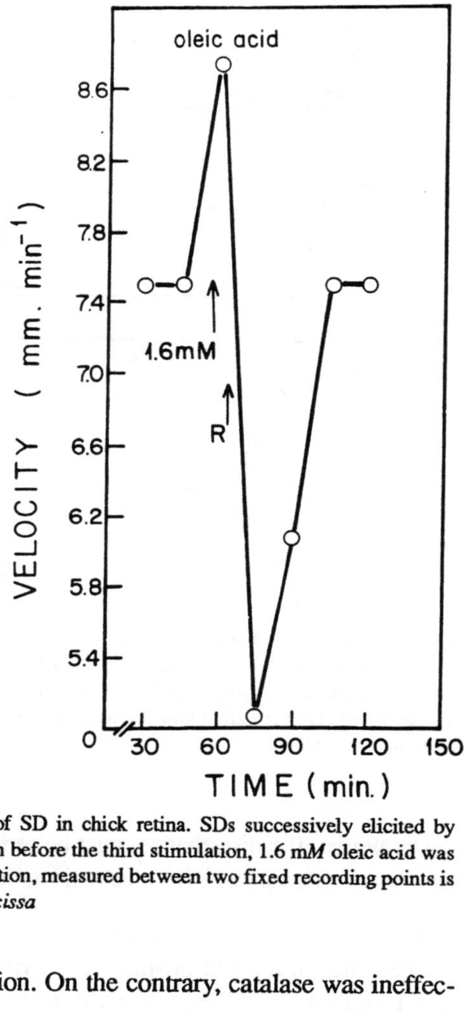

Fig. 7. Effect of oleic acid on the velocity of propagation of SD in chick retina. SDs successively elicited by mechanical stimulation every 15 min. *At the first arrow*, 2 min before the third stimulation, 1.6 m*M* oleic acid was administered. At *R* it was washed out. The velocity of propagation, measured between two fixed recording points is indicated *on the ordinate*; the time of experiment is *on the abscissa*

In some experiments, it could even stop the reaction. On the contrary, catalase was ineffective under the same conditions.

Discussion

Experimental investigation has shown, time and again, that SD is a complex phenomenon. In fact, nearly 50 years after its discovery by Leão (1944), the identification of the cause of its occurrence has proven to be a difficult task. Yet, valuable clues to the understanding of the reaction have been obtained by investigations directed to factors that influence its initiation and propagation. It has been amply demonstrated that stimuli of different sorts (mechanical, chemical, electrical, thermal, etc.) are capable of triggering the reaction. In spite of being diverse in nature, these stimuli drive the preparation to a state which is invariably the same. Disturbing the dynamic equilibrium prevalent in the tissue by a process that at some stage of its evolution ought to be common to them all, the stimuli ultimately bring on the set of events that define SD. As spread is one of the cardinal features of the

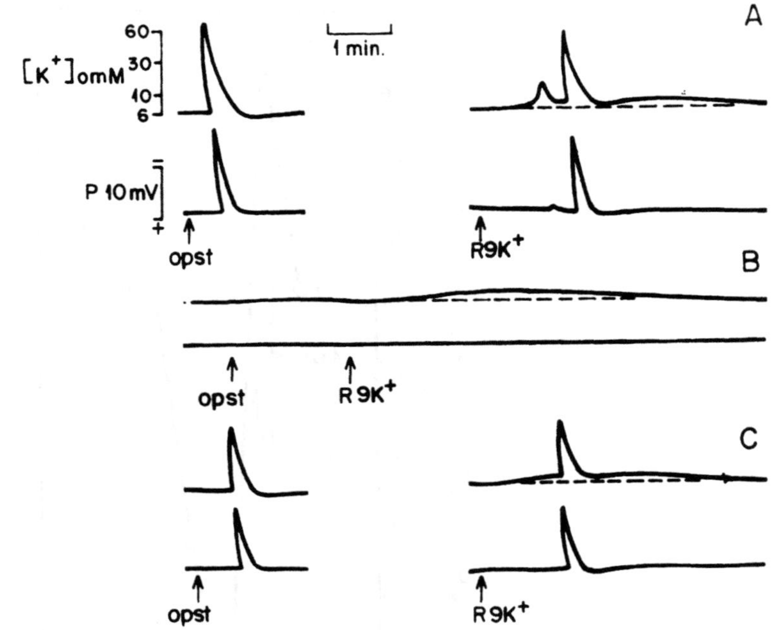

Fig. 8A-C. Effect of chorpromazine on the elicitation of SD by optical and chemical (9 m*M* KCl) stimulation. $[K^+]_o$, extracellular K^+ activity. *P*, slow potential change. A and C show the effects of optical (*op st*) and chemical (*R 9K⁺*) stimulation respectively before and after administration of chlorpromazine. **B** shows the same stimuli applied 10 min after initiating superfusion with Ringer's solution containing 100 m*M* chlorpromazine. Segments of continuous recording

phenomenon, the disturbance must be reproduced anew at the propagating wave front, but now by an autoregenerative mechanism, i.e., independent of the initially applied stimulus.

Among the components that constitute the event of SD, the transient failure of the selective permeability of the cell membrane is striking. This altered membrane function leads to ion translocations to and from the extracellular compartment that do not differ qualitatively from the ones detected by Chan and Fishman (1980, 1985) during brain edema induced by unsaturated fatty acids. Those studies led their authors to propose that cyclic interactions between polyunsaturated fatty acids and oxygen free radicals play a key role in membrane damage, development of edema, and necrosis induced by ischemia and injury.

The conception that free radicals could also be implicated in the generation of SD was first tested by studying the photodynamic action of rose bengal in isolated retina preparations (Netto and Martins-Ferreira 1989), which demonstrated that, under appropriate experimental conditions, SD may be elicited in isolated retina, without irreversible damage, by rose bengal photoactivation.

Although we have no direct evidence for the formation of free radicals in our experimentals, the present results showing that unsaturated fatty acids as applied were able to initiate and under certain conditions increase the velocity of propagation of SD are consonant with the

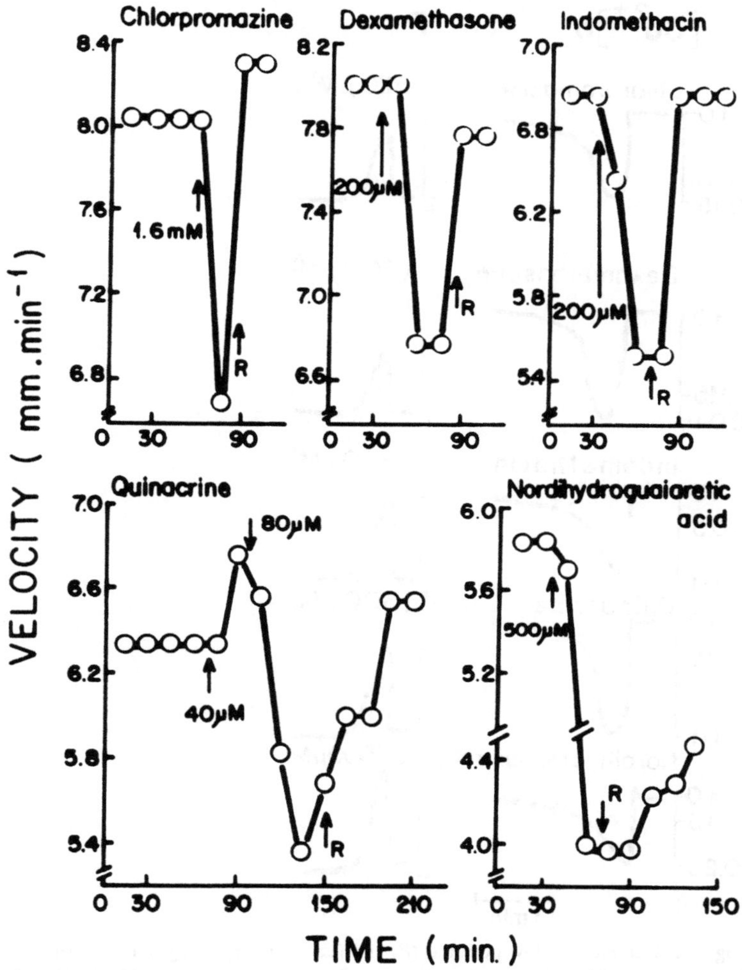

Fig. 9. Effects of substances that interfere with the liberation or with the metabolism of unsaturated fatty acids on the velocity of propagation of SD in chick retina. SDs were elicited by mechanical stimulation every 15 min. *At the first arrow*, the substances were applied at the concentrations indicated *in each panel*. At *R*, the drug was washed out

above concept. At this point, it is appropriate to make a brief comment about our experimental conditions, in connection with the observations of Gutteridge (1988), and Gutteridge and Kerry (1982) on lipid peroxidation. In our experiments no attempt was made to purify the fatty acids utilized which, according to them, probably contained products of lipid peroxidation. So, these products alone could have been responsible for the observed effects.

A notable experimental fact in this study that deserves further investigation is the intensification of the SD-evoking effect of the unsaturated fatty acids in conditions of low Ca^{2+}. Interestingly, the influence of Ca^{2+} on the peroxidation of ox brain liposomas was observed by Gutteridge (1977), who reported that Ca^{2+} both inhibited and enhanced peroxidation in

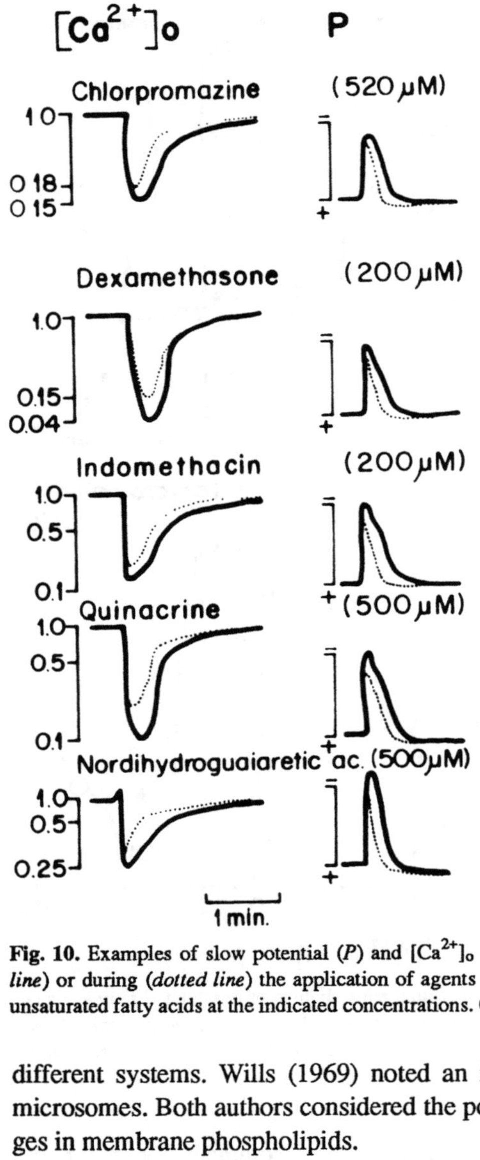

Fig. 10. Examples of slow potential (*P*) and $[Ca^{2+}]_o$ (m*M*) changes accompanying SD either before (*continuous line*) or during (*dotted line*) the application of agents that interfere with the liberation or with the metabolism of unsaturated fatty acids at the indicated concentrations. Calibration in *P* corresponds to 20 mV in all records

different systems. Wills (1969) noted an inhibitory effect of Ca^{2+} on peroxidizing liver microsomes. Both authors considered the possibility of Ca^{2+}-inducing conformational changes in membrane phospholipids.

Our experiments have indicated that at least two alterations were produced in the tissue by administration of the fatty acids. One, early alteration, was characterized by enhancement of the velocity of propagation of SD and increase in amplitude and duration of the slow potential and ionic changes accompanying the reaction. The other, appearing later, is manifest by a retardation in the propagation coinciding with a decrease in amplitude and faster recovery of the slow potential and ionic changes.

If free radicals appearing either during the enzymatic peroxidation of unsaturated fatty acids and/or directly conveyed to the preparation as a product of autoxidation of these acids, are involved in the mechanism of SD, then substances interfering with their liberation or protecting against their action are expected to counteract the reaction. In this rather panoramic view, that prediction was found to be true. Accordingly, chlorpromazine, nor-dihydroguaiaretic acid, and indomethacin raised the threshold for optical, mechanical and chemical stimulation of the phenomenon. These substances, in addition to dexamethasone and quinacrine decreased the velocity of propagation of SD. All of them decreased the amplitude and the duration of the slow potential and ionic changes of the reaction. SOD, which is one of the primary biological defense substances (Fridovich 1988), catalyzing the conversion of superoxide radical to hydrogen peroxide, was also found to oppose SD.

Summary

A number of pathological conditions of the brain have been shown to be associated with the accumulation of unsaturated fatty acids, the metabolism of which involves the formation of free radicals leading to cell membrane alterations. It has also been shown that during the phenomenon of spreading depression (SD) there is a transient breakdown of the selective permeability of the neural cell membrane, and an increase in concentration of arachidonic acid. In addition, SD may be initiated in retina preparations by rose bengal photodynamic action. These facts lead to the suspicion that free radicals are fundamentally involved in the mechanism of the phenomenon. In the present work we made further investigations along these lines by studying the effects of exogenous administration of unsaturated fatty acids on isolated chick retinas. The reaction was elicited by mechanical or chemical stimulation and recorded with single-barreled and double-barreled K^+ and Ca^{2+}-sensitive microelectrodes. The results showed that unsaturated fatty acids trigger SD, alter its velocity of propagation, and slow potential and ionic changes. Moreover, agents that interfere with the liberation or with the metabolism of unsaturated fatty acids oppose the phenomenon.

References

Barkai AI, Bazan NG (eds) (1989) Arachidonic acid metabolism in the nervous system. Ann NY Acad Sci 559

Billah MM, Lapetina EG (1982) Formation of lysophosphatidylinositol in platelets stimulated with thrombin or ionophore. J Biol Chem 257: 5196-5200

Bindoli A, Rigobello MP, Favel A, Galzigna L (1988) Antioxidant action and photosensitizing effects of three different chlorpromazines. J Neurochem 50: 138-141

Blackwell GL, Duncombe WG, Flower RJ, Parsons MF, Vane JR (1977) The distribution and metabolism of arachidonic acid in rabbit platelets during aggregation and its modification by drugs. Br J Pharmacol 59: 353-366

Chan PN, Fishman RA (1980) Transient formation of superoxide radicals in polyunsaturated fatty acid induced brain swelling. J Neurochem 35: 1004-1007

Chan PH, Fishman RA (1985) Free fatty acids, oxygen free radicals, and membrane alterations in brain ischemia and injury. In: Plum F, Pulsinelli W (eds) Cerebrovascular diseases. Raven, New York, pp 161-168

Chan PH, Yurko M, Fishman RA (1982) Phospholipid degradation and cellular edema induced by free radicals in brain cortical slices. J Neurochem 38: 525-531

Demopoulos HB (1972) Molecular aspects of membrane structures in cerebral edema. In: Renhen HJ (ed) Steroids and brain edema. Springer, Berlin Heidelberg New York, pp 29-39

do Carmo RJ, Martins-Ferreira H (1984) Spreading depression of Leão probed with ion-selective microelectrodes in isolated chick retina. An Acad Bras Ciênc 56: 401-421

do Carmo RJ, Martins-Ferreira H (1988) Extracellular Ca^{2+} and retinal spreading depression. Braz J Med Biol Res 21: 607-610

Flower RJ (1974) Drugs which inhibit prostaglandin biosynthesis. Pharmacol Rev 26: 33-67

Flower RJ (1988) Lipocortin and the mechanisms of action of the glucocorticoids. Br J Pharmacol 94: 987-1015

Fridovich I (1988) The biology of oxygen radicals: general concepts, In: Oxygen radicals and tissue injury, Proceedings of an Upjohn Symposium, Barry Halliwell (ed), pp 1-5

Gutteridge JMC (1977) The effect of calcium on phospholipid peroxidation. Biochem Biophys Res Commun, 74: 529-537

Gutteridge JMC (1988) Lipid peroxidation: some problems and concepts. In: Halliwell B (ed) Oxygen radicals and tissue injury. Proceedings of an Upjohn symposium. Federation of the American Societies for Experimental Biology, Bethesda, Maryland, USA

Gutteridge JMC, Kerry PJ (1982) Detection by fluorescence of peroxides and carbonyls in samples of arachidonic acid. Br J Pharmacol 76: 459-461

Hamberg M (1976) On the formation of thromboxane B_2 and 12-L-hydroxy-5,8,10,14-eicosatetraenoic acid (12 ho-20:4) in tissues from guinea pig. Biochim Biophys Acta 431: 651-654

Hirata F, Shiffnan E, Venkatasubramaniam K, Salomon D, Axelrod J (1980) A phospholipase A_2 inhibitory protein in rabbit neutrophils induced by glucocorticoids. Proc Natl Acad Sci USA 73: 1720-1734

Lapetina EG, Billah MM, Cuatrecasas P (1981) The initial action of thrombin on platelets. J Biol Chem 256: 5037-5040

Lauritzen M, Hansen AJ, Kronborg D, Wieloch T (1990) Cortical spreading depression associated with arachidonic acid accumulation and preservation of energy charge. J Cereb Blood Flow Metab 10: 115-122

Leão AAP (1944) Spreading depression of activity in the cerebral cortex. J. Neurophysiol 7: 359-390

Martins-Ferreira H, do Carmo RJ (1987) Retinal Spreading depression and the extracellular milieu. Can J Physiol Pharmacol 65: 1092-1098

Netto M, Martins-Ferreira H (1989) Elicitation of spreading depression by rose bengal photodynamic action. Photochem Photobiol 50: 229-234

Tappel AL, Lundberg WO, Boyer PD (1953) Effect of temperature and antioxidants upon the lipoxidase-catalysed oxidation of sodium linoleate. Arch Biochem Biophys 42: 293-304

Vercesi A, Martins-Ferreira H (1983) Oxygen and glucose requirements in chick retinal spreading depression. An Acad Bras Ciênc 55: 309-316

Wills ED (1969) Lipid peroxide formation in microsomes: the role of non-haem iron. Biochem J 113: 325-332

Magnetic and Impedance Measurements for the Detection of Spreading Depression at a Distance

A. R. Gardner-Medwin

Deptartment of Physiology, University College London, London WC1E 6BT, UK

Introduction

This paper considers two clinical situations where it would be useful to be able to monitor the occurrence of spreading depression (SD) in brain tissue without the use of invasive electrodes or other probes. Study of SD in the retina has the great advantage over studies in the brain that the disturbance is clearly visible as a result of the associated changes of light scattering (Martins-Ferreira and Oliveira-Castro 1966). This has made it possible to identify the initiation, propagation, and failure of SD throughout the whole of the retinal tissue in a manner that otherwise would have required complex, multiple, and potentially traumatic recording probes. The result has been a better understanding of the conditions for propagation (Martins-Ferreira 1983). Nevertheless, the most important advances in the understanding of SD both in the brain and in the retina (see Bureš et al. 1974; Nicholson and Kraig 1981; do Carmo and Martins-Ferreira 1984, for reviews) have been made through microelectrode and chemical studies. Why, then, do we need remote techniques for studying SD in the brain?

The Two Clinical Problems

The two situations in which such techniques are needed both relate to the clinical study of SD in man. The rationale and the technical requirements are very different. The first is for the study of migraine. It has long been thought possible (Leão and Morrison 1945) that SD occurs in classic migraine, i.e., the form of migraine in which there are acute symptoms of neurological dysfunction preceding the headache (for reviews, see Gardner-Medwin 1981; Lauritzen, this volume). Studies on migraine sufferers require totally nontraumatic and preferably noninvasive techniques.

One such technique for the study of migraine has existed for a number of years: the measurement of regional cerebral blood flow (rCBF). This has contributed greatly to the characterization of migraine attacks (Olsen et al. 1981), and in many respects supports the possible involvement of SD in classic migraine (see Hansen and Lauritzen 1984 for review). Unfortunately, however, the circulatory changes are only indirect consequences of SD, and might plausibly result from any other mechanism causing the migraine aura. It is possible (Olsen et al. 1987), though controversial (Kronborg et al. 1990), that blood flow may fall low enough in some patients directly to cause neurological dysfunction. This could be true, however, whether SD occurs or not to trigger the blood flow changes. More direct techniques are required if the occurrence of SD is to be clearly established or refuted. Meanwhile, the evidence that is hardest to explain on any hypothesis other than the SD hypothesis would appear to remain the remarkable quantitative and qualitative similarity that SD bears to the aura in classic visual migraine (Milner 1958).

The second situation where the remote detection of SD would be valuable is in neurosurgery. It is plausible, but not certain, that SD occurs sometimes during neurosurgery in man (Gardner-Medwin and Mutch 1984). The safety factors for metabolic provision are often reduced during surgery, and some of the conditions render SD longer lasting (Mutch and Gardner-Medwin 1987) and a potentially more damaging disturbance than it would be in a normal healthy individual. Blood flow and metabolic provision that are adequate for the healthy survival or normal tissue may result in frank anoxia with the added metabolic load presented by SD (see e.g., Haselgrove et al. 1990). The detection of SD and related phenomena (e.g., local anoxic depolarization) during surgery could represent a valuable clinical tool, both for the identification of possible causes of post-operative complications and for their alleviation if sufficiently rapid warning can be given of their incipient development. The prime requirements for a technique that could be useful in neurosurgery are (1) that it be compatible with practice in the operating theater and (2) that it should not increase the risks of tissue trauma. For these reasons, any probe devices must not penetrate the tissue surface. We are again faced with the problem of detecting SD from a distance if sensitivity is to be maintained for more than purely superficial disturbances.

Nature of the Disturbances during SD

In rat cortex, SD propagates at a uniform rate through fairly flat tissue. This means that a cross-sectional map of the changes within a propagating SD wave can be inferred, at least approximately, from measurements at different depths on a single-electrode track (Fig. 1). The zone of extracellular negativity and high extracellular potassium concentration ($[K^+]_o$) extends from the surface to a depth of 1-2 mm, and is 1 mm or more in width, varying with the ambient conditions, for example blood flow and PCO_2. In the perpendicular direction (parallel to an advancing wave front) the region of tissue affected at any one instant can be much more extensive, commonly forming a complete ring around a focus of initiation in the manner that is seen so clearly in the chick retina preparation (Martins-Ferreira 1983).

The physical changes that might be used to detect the occurrence of SD include those directly associated with the electrophysiological events: the changes of ionic and neurotransmitter concentrations, the prolonged voltage shifts and associated currents, the cessation of action potentials, and the changes of cellular and extracellular volumes. There are also changes that arise in a secondary manner as physiological responses to the primary events of SD, in at least some circumstances: changes in active transport, metabolic biochemistry, blood flow, glucose uptake, oxygen availability, etc. In many of these respects, SD produces larger

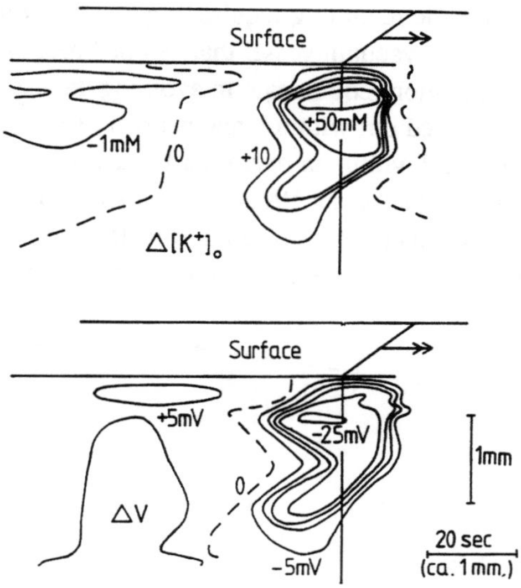

Fig. 1. Contour plots showing the region of tissue affected by extracellular K^+ changes (*above*) and extracellular voltage changes (*beneath*) during SD in parietal cortex of a urethane-anesthetized rat. The plots were derived from K^+ microelectrode recordings at different depths on one electrode track, during eight successive propagating waves of SD at 6 min intervals. Records were aligned in time relative to the peak negativity recorded simultaneously from a fixed Ag/AgCl electrode at the surface (*vertical line*). The *horizontal axis* is strictly a time axis, but is equivalent to a horizontal spatial axis (with calibration matched to the vertical, depth axis) for propagation at 3 mm/min (*toward the right*) through uniform flat tissue. Data obtained in collaboration with J. L. Skelton

changes than are known to occur in the course of normal physiological activity.

Very many techniques have been used to observe these changes, either directly or indirectly. I shall examine the potential of two specific techniques for assisting with the particular clinical problems outlined above: magnetic field measurement in relation to the detection of SD in neurological disturbances such as migraine, and electrical impedance measurement in relation to the detection of SD and anoxic regions beneath the tissue surface in neurosurgery.

Magnetic Fields due to SD

In some circumstances the current flow around a region of SD can produce distance voltage changes and magnetic fields. The voltage changes are restricted to the tissue, and are both attenuated and smoothed over a large area on the scalp by the combined effect of the high resistance of the meninges and skull and the relatively low resistance of the scalp. The magnetic fields extend into surrounding space and are not attenuated by the skull and the scalp. Magnetic recordings during SD were first made by Okada et al. (1988), using an in vitro preparation from the turtle cerebellum. Recordings have also been made in vivo during SD in the cortex of intact anesthetized rabbits (Gardner-Medwin et al. 1990, 1991). The fields are small [a few picoTesla (pT) at 20 mm distance, cf. the Earth's magnetic field, ca. 50 μT].

Magnetic recording requires superconducting sensor coils immersed in liquid helium within a Dewar vacuum vessel that can be brought close to the head (Fig. 2). The tissue and animal preparation require care in avoiding the use of magnetic materials, and the experiments must be carried out within a magnetically screened chamber. The technique is unlikely ever to be practical within an operating theater, but it can be used with migraine subjects during an attack (Tepley et al. 1990). It is possible to record simultaneously from multiple coils, and one of the advantages in relation to the study of neurological patients is that no electrodes or other attachments to the patient are required, and hence there is very little delay in com-

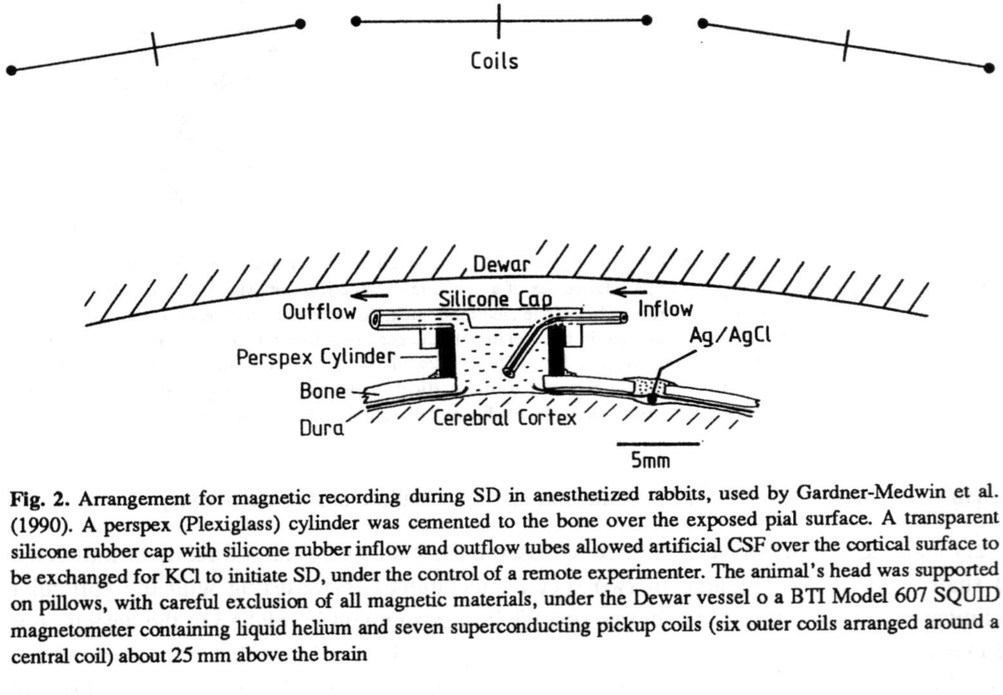

Fig. 2. Arrangement for magnetic recording during SD in anesthetized rabbits, used by Gardner-Medwin et al. (1990). A perspex (Plexiglass) cylinder was cemented to the bone over the exposed pial surface. A transparent silicone rubber cap with silicone rubber inflow and outflow tubes allowed artificial CSF over the cortical surface to be exchanged for KCl to initiate SD, under the control of a remote experimenter. The animal's head was supported on pillows, with careful exclusion of all magnetic materials, under the Dewar vessel o a BTI Model 607 SQUID magnetometer containing liquid helium and seven superconducting pickup coils (six outer coils arranged around a central coil) about 25 mm above the brain

mencing a recording session. Against this must be set the need for careful control for sources of magnetic interference that can lead to artifacts due to magnetic contamination and patient movement. For example, dental materials and hair spray can both cause problems that can vary between sessions even with the same subject.

Typical signals recorded during SD propagation in the rabbit cortex are shown in Fig. 3 (Gardner-Medwin et al. 1991). SD was initiated with 150 m*M* KCl in the left parietal cortex, and the trace shows the mean changes of vertical magnetic field for up to 15 min after SD initiation, recorded from coils in one of the sites (posterior and right from the KCl chamber) that gave the largest changes. The consistent component of the magnetic response was a slowly developing upward deflection (upward field through the coil) between about 200 and 600 s after KCl application, with a peak at about 350 s. In addition there were transient downward shifts at the times of KCl application and washout, consistently present in some preparations and absent in others and not statistically significant overall: their origin is not

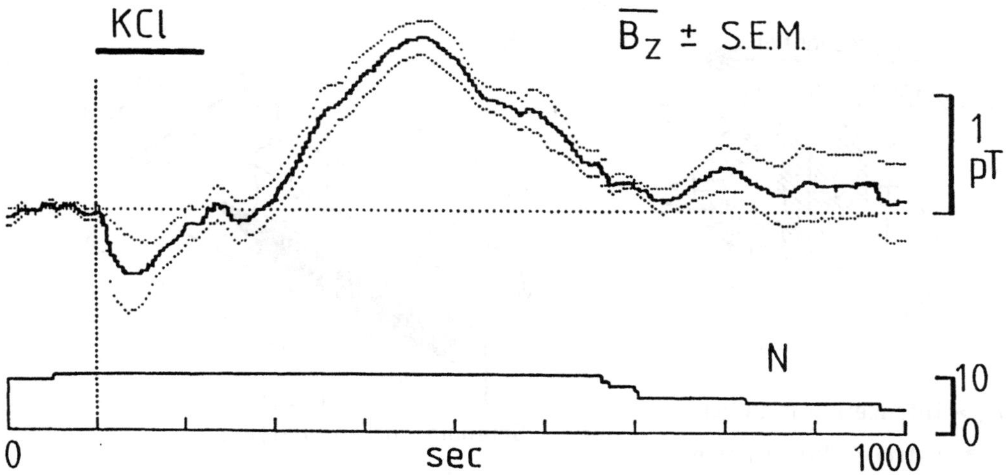

Fig. 3. Mean shifts of magnetic field associated with SD. Initiation of SD was by application of 150 m*M* KCl for 120 s over the left parietal cortex. The mean vertical field (*Bz*) is shown (± 1 SEM) for ten episodes of SD in five animals, for the coil in each recording session that was both to the right and posterior from the initiation site. Records were aligned relative to the start of KCl application. Since different records extended for different lengths of time, the number (*N*) for each part of the trace is plotted beneath. Data from the study by Gardner-Medwin et al. (1990)

known, but may have been connected with the mode of perfusion of the cortical chamber.

In addition to these changes in the DC magnetic recordings, clear changes in the level of magnetic fluctuations in the 1- to 50-Hz bandwidth were occasionally seen (Gardner-Medwin et al. 1990). For technical reasons, these only became evident in analysis of the recorded data after the end of the study, and the conditions for their generation were not investigated. Drops in the level of fluctuations of up to 70% occurred sometimes over a few seconds, sometimes at the time of KCl application and sometimes at other times without any evident precipitating stimulus. It seems more likely that this was related to changes of diffuse activating influences within the nervous system than to the generation of SD.

Mechanisms of Generation of Magnetic Fields in SD

The relations between observed magnetic signals and the details of SD propagation in rabbit cortex are discussed elsewhere (Gardner-Medwin et al. 1991). The general mechanisms by which SD can generate magnetic fields are described here (Fig. 4). The relationship between currents in neural tissue and the magnetic fields they generate is fairly complex (Williamson and Kaufman 1987). There are many patterns of current that lead to electric fields but no magnetic fields at all, for example (in a simple spherical model of the brain and skull) sources of inward and outward membrane current that lie on a line perpendicular to the skull. In general it is necessary for there to be some break in the symmetry of current patterns about a line or a plane perpendicular to the skull, if magnetic fields are to be generated.

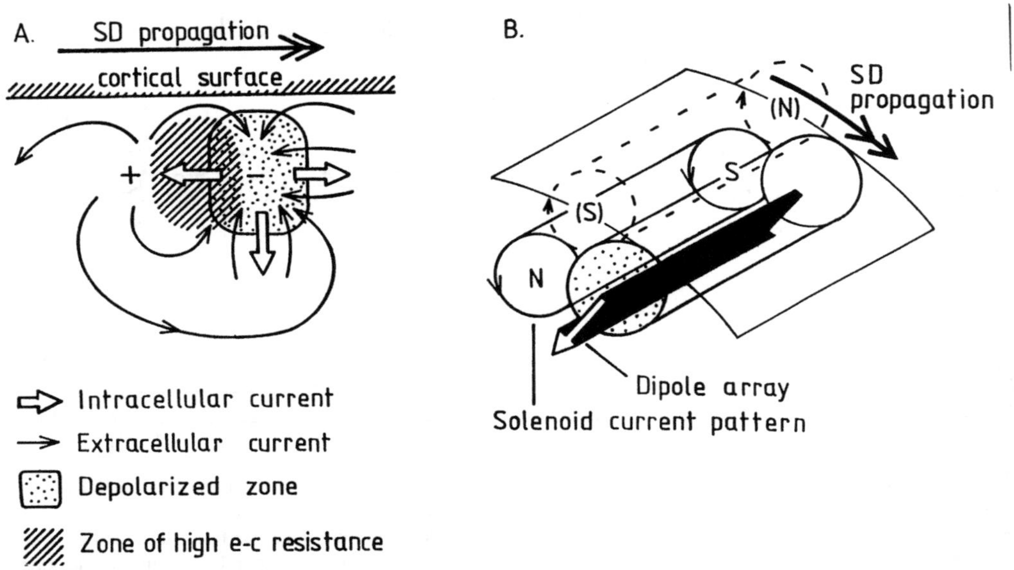

Fig. 4. Diagram of currents (*A*) and principal magnetic generators (*B*) around a region of SD propagating to the right. The depolarized region is shown *stippled* (cf. the zone of extracellular negativity and high $[K^+]_o$ in Fig. 1). The principal intracellular currents (*A*) are away from this region: forward, backward, and toward deeper tissue. The relatively small distance over which intracellular currents can flow toward the surface leads to a net intracellular dipole array toward deeper tissue (*B*). In addition, there is asymmetry between the extracellular resistance around the zones of forward and backward currents, due to the high extracellular resistance caused by SD (*hatched zone*). This leads to more of the backward current returning via long-distance extracellular current loops, rather than through the local extracellular space. If the cortical surface lies against high-resistance tissue such as dura and bone, these loops are predominantly in the deep tissue and have a magnetic effect similar to that of a solenoid (*B*), with a north pole (*N*) to the right of a restricted zone of propagation and a south pole (*S*) to the left. The solenoidal currents have no magnetic effect if the SD wave forms a complete ring around a point of initiation

Figure 4A shows the principal intracellular current flow associated with SD (hollow arrows). This is probably largely glial current associated with K^+ clearance (Gardner-Medwin 1981). These arrows correspond to current dipoles, with outward membrane current at their heads and inward current at their tails. The magnetic effect of the dipoles can be calculated within uniformly resistive tissue without considering the return extracellular current paths, since by symmetry the net effect of these return paths is zero. Since the forward and backward dipoles associated with SD are similar and largely cancel each other out in their magnetic effects, the principal magnetically effective intracellular dipole is away from the adjacent tissue surface (Okada et al. 1988). Thus one of the principal ways in which SD can generate a magnetic field is through an array of dipoles pointing away from the cortical surface throughout the zone of depolarization (Fig. 4B). A field is only generated by this mechanism, however, where these dipoles are not perpendicular to the skull, i.e., where SD propagates in cortex facing a sulcus such as the sagittal sulcus in the rabbit, or the numerous sulci in the convoluted cortex of man.

A second asymmetry of the current flow, capable of generating a magnetic field, applies also to SD propagating parallel to the skull. This results from an increase of extracellular impedance that is more pronounced in the wake of SD (Hoffman et al. 1973), due to cellular NaCl influx and reduced extracellular ion content (Fig. 4A). Much of the intracellular current flowing in front of the zone of SD depolarization returns in the adjacent extracellular space, thereby cancelling its magnetic effect. This return pathway is partially occluded for currents behind the zone of depolarization, which therefore follow long looping extracellular pathways (Fig. 4A). If the tissue surface lies against a zone of high resistance (e.g., the skull and meninges), then the long extracellular paths are predominantly in deeper tissue, generating a solenoidal pattern of current flow with N and S poles respectively on the right and left of a propagating SD wave front (Fig. 4B). If the wave of SD forms a complete ring around an initiating focus, then the solenoidal current pattern is closed in a torus shape, and there is no magnetic effect. Thus this second mechanism generates magnetic fields only where the SD wave front is broken through failure of propagation at some point: a condition that applies in the retrosplenial cortical area in the rabbit (Leão 1944) and presumably in certain areas in man.

Implications for Magnetic Studies in Migraine

Various lessons can be learned from the animal studies, relevant to the characterization of neurological disturbances in migraine or other disorders. The signals measured during SD in the rabbit were very slowly changing (Fig. 3), close to the limits of baseline stability for their detection in individual records and requiring DC recording conditions. Most prior records made with migraine patients had been carried out with AC recording (Tepley et al. 1990). The signals from the rabbits were smaller and slower to change than those previously observed with SD in vitro (Okada et al. 1988). The conditions in vitro are especially favorable for magnetic recording in two respects: the tissue was surrounded by a saline solution with a substantially higher conductivity than the brain tissue around a zone of SD in vivo. This tends to reduce voltage changes and increase magnetic changes. SD was also initiated with electrical stimulation, producing synchronous SD over a relatively large area.

Even the type of signals detected in the rabbit could in principle be identified if they occur in man associated with migraine attacks, especially if the subject is constantly kept moving under the magnetometer (Gardner-Medwin 1991). It is also possible that larger signals than those seen in the rabbit might be recorded if SD does occur in migraine, because of the large amount of tissue that may be involved and its potentially more favorable orientation. Such signals appear well worth seeking. The magnetic signals are likely to be highly dependent on the shape and orientation of the zones affected by SD. It is clear therefore that even large signals may be difficult to characterize as definitely due to SD rather than to zones of ischemia or other disturbances that might account for the symptoms of migraine.

Use of Impedance Techniques for Detecting Deep Tissues Disturbances

The problem of detecting SD during neurosurgery is twofold: that of getting adequate sensitivity to disturbances beneath the tissue surface, with techniques limited to surface access, and that of distinguishing signals arising from deep disturbances from those arising nearer the surface. Impedance measurements offer a good prospect of achieving these goals, within limits that have yet to be determined.

It might be asked why it is necessary to distinguish surface disturbances from deep disturbances if both can be detected and both are known to be undesirable. The problem is that most techniques based on measurements at the surface will be more sensitive to superficial than to deep disturbances. Thus the source of a given signal may in general be either a mild superficial disturbance or a more severe deep disturbance, with possibly quite distinct clinical implications.

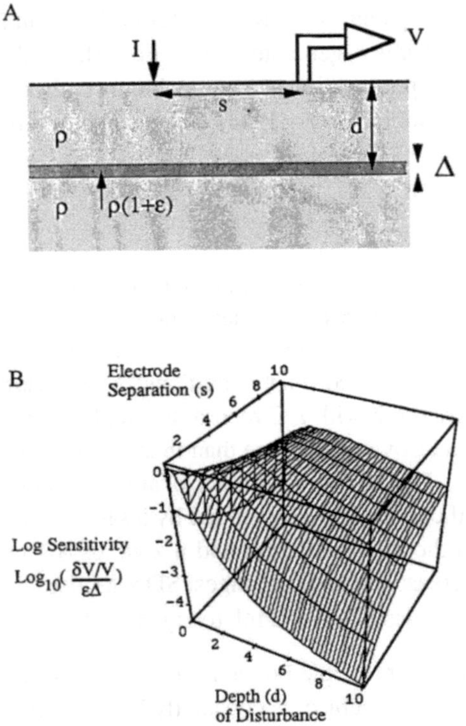

Fig. 5A,B. Detection of impedance disturbances under the tissue surface with surface current application and voltage measurements. **A** A zone of tissue of thickness Δ at depth *d* has its resistivity (ρ) raised by a fraction ε. Current is applied at a point on the surface (I), with a remote return electrode. The voltage gradient at the surface (*V*) is measured with a pair of electrodes at a distance *s* from the current injection site. The increase of ρ leads to a fractional increase $\delta\, V/V$, proportional (for relatively small changes) to $\varepsilon\,\Delta$. The sensitivity (**B**) is defined as the ratio of these quantities and is plotted on a logarithmic scale as a function of both *d* and *s*. At short *s* it falls rapidly with depth *d*, while with large *s* the sensitivity increases with depth (up to $d = s/\sqrt{6}$) and then falls more slowly

In principle impedance measurements or "applied potential tomography" (Brown et al. 1985) can generate enough information to produce a complete image of disturbances of impedance within a tissue from just surface measurements, provided there are no zones surrounded by complete insulators. The present requirements fall far short of the need for images, and may perhaps be realized with much less than the optimal conditions.

One situation in which surface measurements can yield the necessary information about the depth of an impedance disturbance is illustrated in Fig. 5A. Current (*I*) is passed from a surface electrode over the tissue of interest to a remote return electrode. The resulting voltage gradient (*V*) is measured with a pair of surface electrodes at a distance *s*. The sensitivity of *V* to a widespread change of impedance at a specific depth *d* is plotted in Fig. 5B. With a small electrode separation *s* the sensitivity to impedance changes falls rapidly with depth. An increase of the electrode separation lowers the sensitivity to superficial disturbances but raises the sensitivity to deep disturbances. Thus by observing the ratio of changes measured with different separations it is possible to deduce the depth of the disturbance.

This example illustrates the general principle of inferring the depth of a disturbance. Of course disturbances in general are not uniform at a constant depth in the simple manner assumed above. However, by adding measurements made with many current injection sites as well as many sites of recording, it is possible to deduce the full pattern of impedance changes in both depth and horizontal position: the principle of applied potential tomography. The depth and resolution to which such information can be obtained depends on the number of measurement sites, their separation, and the noise level of the measurements.

Impedance Measurements during SD

SD has long been known to be associated with impedance increases detectable from surface measurements (Leão and Martins-Ferreira 1953). The local increases beneath the surface are as great as a factor of 2-3 (Hoffman et al. 1973). Holder and Gardner-Medwin (1988) have made surface measurements under conditions that permit a comparison of the signals with those expected for disturbances at specific depths (Fig. 6). Alternating current (50 kHz) was passed between the first and last of an array of nine electrodes (Fig. 6A). The in phase amplitude of the resulting 50-kHz potential changes was measured at each of the intermediate electrode sites during passage of an SD wave under the electrode array. In Fig. 6A the changes of this signal from baseline values are plotted against time for each electrode site.

The recorded changes in Fig. 6A are largest at the electrodes adjacent to the two current passage sites, and are opposite in sign because of the opposite phase of the current at the two sites. The signal at the far end (b) is also displaced in time because of the propagation delay for the SD traveling between the two sites. The peak changes at the two ends were respectively 46% and 47% of the baseline 50-kHz signals. The depth information can be seen by comparing the signals in Fig. 6A with those calculated for small propagating disturbances at depths of 0.5 mm (Fig. 6B) and 2 mm (Fig. 6C). The changes associated with the smaller depth would fall off more rapidly with distance between recording and current passage, while for the greater depth the signals would fall off less rapidly than the observed data. This is consistent with the range of depths of the SD disturbance in the rat (Fig. 1).

The recordings made from the different electrodes for Fig. 6A were made sequentially during repeated SD episodes. It is possible, however, to multiplex signals from such

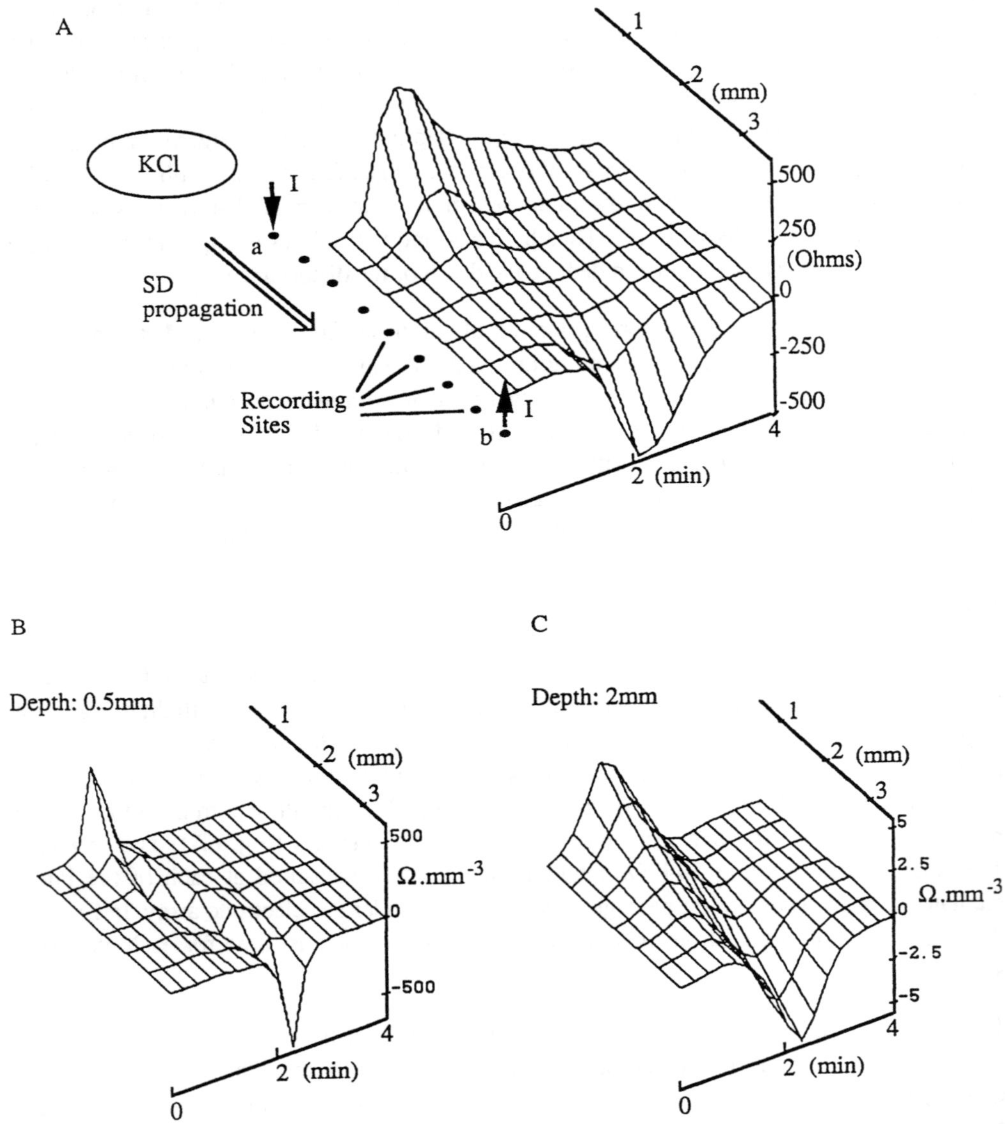

Fig. 6A-C. Impedance measurements during SD in the cortex of an anesthetized rat. **A** SD was initiated with 150 m*M* KCl in a cup (marked KCl) and propagated along a row of nine surface electrodes (a-b) at 0.5-mm spacing. Alternating current was passed between the first and the last electrodes (I, ca. 20 μA p/p, 50 kHz). The voltage changes at 50 kHz (relative to a remote earth) were selectively filtered and their in phase components measured with a phase-locked detector during successive SD waves at each of the intermediate electrodes. The changes during SD are plotted as functions of time adjacent to each recording site. They are expressed in ohms, as the voltage change divided by the applied current. **B,C** show the calculated signals expected for a small zone with a high impedance increase propagating at depths of 0.5 mm (**B**) and 2 mm (**C**) beneath the electrodes. These are plotted in the same way as **A**, and expressed per cubic millimeter of tissue rendered nonconductive. Baseline resistivity = 420 Ω. (Data and calculations redrawn from Holder and Gardner-Medwin, 1988)

electrodes with a single detector circuit (Brown et al. 1985) so that all the necessary measurements could be carried out essentially simultaneously on a single run. It is evident that the same principle explained above for widespread disturbances (Fig. 5), whereby the recorded signals are more strongly dependent on electrode separation for superficial than for deep disturbances, applies also for small zones of disturbance (Fig. 6B, C).

The optimal electrode configurations and separations for impedance monitoring during surgery can be tailored to specific applications such as tissue particularly at risk beneath retractors (Gardner-Medwin and Mutch 1984). The limits of resolution of such techniques deserve to be determined.

Summary

Two distinct reasons are set out for attempting to detect spreading depression (SD) from a distance (i.e., through measurements at remote sites): (1) the need for aditional evidence relevant to the possible occurence of SD in migraine and (2) the advisability of monitoring for the occurrence of SD and allied phenomena during surgery.

Biomagnetic measurements may be able to detect SD non-invasively in man, but the nature of the signals obtained from anesthetized rabbits during propagating SD suggests that improved stability of DC recording techniques in man may be required. When there is neurosurgical exposure of vulnerable brain tissue, the monitoring of SD and anoxia beneath the surface, causing local changes of electrical impedance, may be carried out through the application of current and the measurement of voltage changes restricted to the tissue surface. In neurosurgical monitoring an important distinction must be made between mild superficial disturbances and deeper profound disturbances, which in some respects produce similar signals at the surface. Information about the depth of the disturbance can be extracted from multiple surface impedance measurements, and may assist in making this distinction.

References

Brown BH, Barber DC, Seagar AD (1985) Applied potential tomography: possible clinical applications. Clin Phys Physiol Meas 6: 109-121

Bureš J, Burešová O, Křivánek J (1974) The mechanism and applications of Leão's spreading depression of electroencephalographic activity. Academia, Prague

do Carmo RJ, Martins-Ferreira H (1984) Spreading depression of Leão probed with ion selective microelectrodes in chick retina. An Acad Brasil Ciênc 56(4): 401-421

Gardner-Medwin AR (1981) Possible roles of vertebrate neuroglia in potassium dynamics, spreading depression and migraine. J Exp Biol 95: 111-127

Gardner-Medwin AR (1991) Rotatory scanning for improved discrimination of very low frequency components of the magnetoencephalogram. Proceedings of the 8th International Biomagnetism Conference. Münster, FRG

Gardner-Medwin AR, Mutch WAC (1984) Experiments on spreading depression and neurosurgery. An Acad Brasil Ciênc 56(4): 423-430

Gardner-Medwin AR, Tepley N, Barkley GL, Moran J, Nagel-Leiby S, Simkins RT, Welch KMA (1990) Magnetic observation of spreading cortical depression in anaesthetized rabbits. In: Williamson SJ et al. (eds) Advances in biomagnetism. Plenum, New York, pp 323-326

Gardner-Medwin AR, Tepley N, Barkley GL, Moran J, Nagel-Leiby S, Simkins RT, Welch KMA (1991) Magnetic fields associated with spreading depression in anaesthetised rabbits. Brain Res 540: 153-158

Hansen AJ, Lauritzen M (1984) The role of spreading depression in acute brain disorders. An Acad Brasil Ciênc 56(4): 457-479

Haselgrove JC, Bashford CL, Barlow CH, Quistorff B, Chance B, Mayevsky A (1990) Time resolved 3-dimensional recording of redox ratio during spreading depression in gerbil brain. Brain Res 506: 109-114

Hofmann CJ, Clark FJ, Ochs S (1973) Intracortical impedance changes during spreading depression. J Neurobiol 4: 471-486

Holder DS, Gardner-Medwin AR (1988) Some possible applications of applied potential tomography. Clin Phys Physiol Meas 9 [Suppl A]: 111-119

Kronborg D, Dalgaard P, Lauritzen M (1990) Letter to the editor. Arch Neurol 47: 124-125

Lauritzen M, Olesen J (1984) Regional blood flow during migraine attacks by Xe-133 inhalation and emission tomography. Brain 107: 447-461

Leão AAP (1944) Spreading depression of activity in the cerebral cortex. J Neurophysiol 7: 359-390

Leão AAP, Martins-Ferreira H (1953) Alteração da impedância elétrica no decurso da depressão alastrante da atividade do cortex cerebral. An Acad Bras Ciênc 25: 259-266

Leão AAP, Morrison RS (1945) Propagation of spreading cortical depression. J Neurophysiol 8: 33-45

Martins-Ferreira H (1983) Spreading depression in the chick retina. In: Takanoori Ookawa (ed) Brain and behaviour of the fowl. International Specialized BK Services, Portland, pp 317-333

Martins-Ferreira H, Oliveira-Castro G (1966) Light scattering changes accompanying spreading depression in isolated retina. J Neurophysiol 29: 715-726

Milner PM (1958) Note on a possible correspondence between the scotomas of migraine and spreading depression of Leão. EEG Clin. Neurophysiology 10: 705

Mutch WAC, Gardner-Medwin AR (1987) The influence of hypocarbia on the resolution of transient increases in brain extracellular potassium. Anesthesiology 66: 305-355

Nicholson C, Kraig RP (1981) The behavior of extracellular ions during spreading depression. In: Zeuthen T (ed) The application of ion-selective electrodes. Elsevier, Amsterdam, pp 217-238

Okada YC, Lauritzen M, Nicholson C (1988) Magnetic field changes associated with spreading depression: a model for the detection of migraine. Brain Res 442: 185-190

Olesen J, Larsen B, Lauritzen M (1981) Focal hyperemia followed by spreading oligemia and impaired activation of rCBF in classic migraine. Ann Neurol 9: 344-352

Olsen TS, Friberg L, Lassen NA (1987) Ischemia may be the primary cause of the neurologic deficits in classic migraine. Arch Neurol 44: 156-161

Tepley N, Barkley GL, Moran J, Simkins RT, Welch KMA (1990) Observation of spreading cortical depression in migraine patients. In: Williamson SJ et al. (ed) Advances in biomagnetism. Plenum, New York, pp 327-330

Williamson SJ, Kaufman L (1987) Analysis of neuromagnetic signals. In: Gevins AS, Remond A (eds) Handbook of electroencephalography and clinical neurophysiology, vol 1. Elsevier, Amsterdam, pp 405-448

Interstitial Fluid pH and its Changes During Spreading Depression in Isolated Chick Retina

C. R. Ferreira-Filho[1] **and H. Martins-Ferreira**[2]

[1]Departamento de Fisiologia, Instituto Biomédico da Universidade Federal Fluminense, Rua Prof. Ernani Mello 101, 24210, Niterói, RJ, Brazil
[2]Departamento de Neurobiologia, Instituto de Biofísica Carlos Chagas Filho, Universidade Federal do Rio de Janeiro, Centro de Ciências da Saúde – Bl. G, Ilha do Fundão, 21941, Rio de Janeiro, RJ, Brazil

Introduction

The advent of the double-barreled liquid ion-selective microelectrode for sensing H^+ (H^+–ISM) has greatly facilitated the study of brain H^+ activity due to its high selectivity and rapid rise time (Ammann et al. 1981). With this kind of ISM, intra- and extracellular brain pH have been determined and the changes occurring during physiological stimulations as well as some pathological conditions, such as Leão's spreading depression (SD), have been reported in both in vivo and in vitro preparations (Kraig et al. 1983; Mutch and Hansen 1984; Nicholson et al. 1985; Chesler and Chan 1988; Rice and Nicholson 1988).

For a long time the isolated retina has been used in our laboratory as an experimental model for the study of Leão's SD in excised nervous tissue maintained in solutions of known chemical composition (Martins-Ferreira 1962). Reviews of the various aspects of SD in isolated retina have been published recently (Martins-Ferreira 1983; Martins-Ferreira and do Carmo 1987). Voltage and ionic changes that accompany SD in the retina in vitro have been measured with extracellular microelectrodes containing liquid ion-exchanger resins selective for Na^+, K^+, Cl^-, and Ca^{2+} (do Carmo and Martins-Ferreira 1984).

In this study, which has already been reported briefly (Ferreira Filho and Martins-Ferreira 1983; 1985), we have employed H^+–ISMs to measure intraretinal extracellular pH (pH_o) of isolated retinas superfused with solutions of known chemical composition, as well as the pH_o shifts in extracellular space which occur during SD. We have also examined the effects of changing Cl^- concentration in Ringer's solution, inhibitors of carbonic anhydrase, inhibitors of the Na^+/H^+ or Cl^-/HCO_3^- antiporter, and impairment of energy supply both on resting pH_o and on pH_o shifts that accompany SD. The aim of this study was to assess how each of these pH-regulating mechanisms is activated to maintain pH_o homeostasis both in the steady state as well as during SD.

Experimental Procedures

Experiments were performed on white leghorn chicks, 1-7 days after hatching. Birds were decapitated and the eyeballs removed and sectioned along the equator. The vitreous humor was drawn off and the posterior eyecup was placed in a petri dish containing "standard" Ringer's solution (RSt) with the following chemical composition (m*M*): NaCl 100, KCl 6, $MgSO_4$ 1, NaH_2PO_4 1, $NaHCO_3$ 30, glucose 30, $CaCl_2$ 1. RS osmolarity was 290 mOsm. With the exception of the experiments involving gas substitution, the solution was gassed with O_2, and its pH was about 8.9. This solution was chosen because it allows the SD phenomenon to be regularly evoked by optical stimulation.

The eyecup was placed in the hollow receptacle of a plastic chamber partially covered with two glass coverslips. Experiments were usually performed by superfusing the retina at a flow rate of 1 ml/min using a peristaltic pump. The temperature of the bath was controlled with a thermostatic system and, unless otherwise stated, it was kept constant at about 30°C.

Low chloride solutions were made by substitution of sodium chloride with sodium isethionate. Sodium fluoride (2.0 m*M*) or sodium oxamate (1.0 - 10.0 m*M*) were added to block glycolysis; acetazolamide (4.0 - 8.0 m*M*) to inhibit the activity of carbonic anhydrase; amiloride (0.02 - 0.2 mM) to inhibit Na^+/H^+ antiporter; and 4,4'-diisothiocyanostilbene-2,2'-disulfonic acid (DIDS) (0.02 - 0.2 m*M*) to inhibit Cl^-/HCO_3^- antiporter.

The preparation was illuminated with a microscope lamp and the reaction was elicited optically every 15 min by flickering the light, or by mechanical stimulation. The preparation was continuously observed through a dissection microscope and its viability was assessed according to the following criteria: its transparency, its susceptibility to initiation of SD after an adequate stimulus, and the uniform velocity of spread of the wave in all directions.

Double-barreled ISMs were constructed with theta capillary borosilicate glass by the method of Lux (1974) modified as described elsewhere (do Carmo and Martins-Ferreira 1984). The silanized side of the microelectrode was made H^+ selective with the neutral carrier H^+-LIX and was blackfilled with a solution containing 40 m*M* KH_2PO_4, 23 m*M* NaOH, and 15 m*M* NaCl. The ISMs were mounted on a microdrive and connected to a two-channel electrometer (WPI-FD223), and the signals were displayed on a Grass recorder. One channel recorded the potential picked up by the reference barrel of the ISM. The other channel recorded the signal obtained by subtracting the potential picked up by the reference barrel from the potential generated in the H^+-LIX side of the ISM. The indifferent macroelectrode containing Ag/AgCl 1*M* KCl/4% agar was located in the pool of Ringer's solution draining out of the plastic chamber. ISMs were calibrated before and at the end of experiments by applying a bolus of phosphate buffer (pH = 6.0, 7.0, 8.0, 9.0) in the stream of Ringer's solution, the values being checked against a laboratory pH-meter (Beckman-Expandomatic SS2).

All experiments were performed by displacing microelectrodes radially through retinal layers from the vitreal surface. When retinas were superfused with RSt, the records were made 0, 30, 60, 90, 120, 150, and 180 μm beneath the surface from the vitreous side, as measured with a microdrive device, and the results obtained were correlated with the depth profiles of the field potentials reported by do Carmo and Martins-Ferreira (1984). In experiments made by lowering the bath temperature, the microelectrode tips were located at depths of 100 and 180 μm. In all other situations, microelectrode tips were located 100 μm beneath the vitreal surface.

Materials. H^+-LIX neutral carrier was a generous gift from Prof. W. Simon (Swiss Institute of Technology, Zurich). Test drugs and substances were purchased from Sigma (St Louis, Mo., USA) (sodium fluoride, sodium oxamate, acetazolamide, DIDS) and Merck (USA) (amiloride).

Results

Superfusion with Standard Ringer's Solution

Steady State Retinal pH_o. When retinas were superfused with Ringer's solution saturated with 100% O_2 (RSt), the intraretinal steady state baselin pH at the inner plexiform layer (IPL) was 7.16 ± 0.13 (mean ± SD, $n = 45$) while the pH of the Ringer's solution was about 8.9. Although having a pH far beyond that reported for extracellular fluid, this solution was chosen because preparations remained in a satisfactory condition for several hours of experiments; they remained fully transparent in the absence of SD (see Martins-Ferreira 1983), and steady state baseline pH_o was almost the same throughout the experiment. There was a pH gradient between the different retinal layers, the tissue becoming more acidic as the ISM advanced to deeper layers. When the ISM was located in the pool of Ringer's solution at distances less than 200 μm from the retinal surface, a pH gradient in superfusate was also detected as the probe advanced toward the tissue. This disappeared when the solution was stirred (do Carmo et al. 1985). Some experiments were performed by superfusing preparations with solutions gassed with 95% O_2 + 5% CO_2 (pH = 7.5). The steady state baseline pH_o at the inner plexiform layer was more acidic than in control situations (0.1–0.2 units) and the gradient was still present, but preparations became less responsive to optical stimuli.

pH_o Changes During Spreading Depression. The pH_o changes accompanying SD were biphasic in almost all layers of the retina, and showed an initial rapid alkaline shift followed by a smaller and longer acidic shift. The amplitude of these changes depended on the depth of penetration (Fig. 1).

The alkaline shift was detected only when the microelectrode touched the retinal surface. It started concomitantly with the slow voltage variation of the baseline potential (PC) and light-scattering changes (LSCs), and its peak was attained in a few seconds. Its amplitude increased as the ISM was advanced into the retina, and maximal changes were observed at 90-110 μm depth (ΔpH_o = 0.21 ± 0.09, n = 15). As the ISM was advanced into deeper layers, the alkaline shift became smaller; it was not detected at 190 μm depth. Its duration

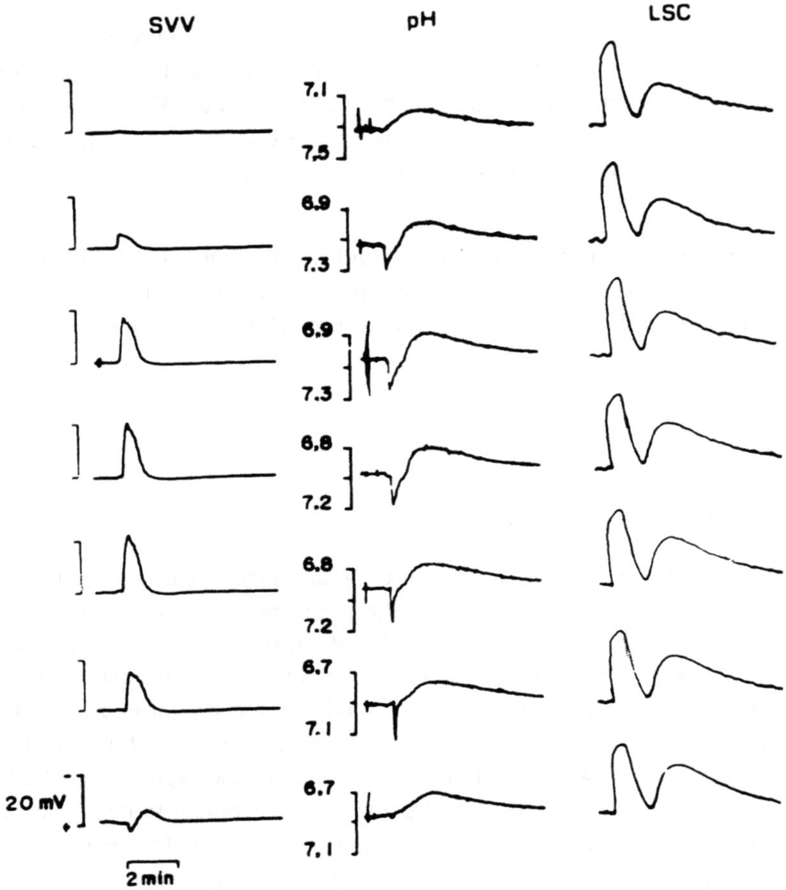

Fig. 1. Depth profile of intraretinal potential (*P*) and pH_o (*pH*) changes during SD, with ISM directed from the vitreous to the receptor side. Depths (*up to down*): 0, 30, 60, 90, 120, 150, and 180 µm. *In the right column* the correspondent changes of light scattered on the retina surface (*LSC*) are also shown

also depended on the depth, being shorter in the deeper layers.

The acidic shift was detected even when the microelectrode tip was located in the superfusate near the retinal surface. Its amplitude became higher and higher as the ISM was advanced into the tissue, attaining its maximum at 90-130 µm depth ($\Delta pH_o = 0.13 \pm 0.04$, $n = 15$). As the ISM was advanced into deeper layers, this shift became smaller and finally vanished. This shift had a total duration of about 6 min, and the peak was attained after the PC had returned to baseline values.

Chloride Substitution

Replacement of Cl^- by isethionate up to 75% is known to increase the susceptibility of the retina for the occurrence of spontaneous waves of SD, with a decrease in the potential changes recorded at IPL (do Carmo and Martins-Ferreira 1984). This experimental procedure caused a resting tissue alkalinization and, when the preparation was again superfused with RSt, the resting pH_o undershot the control values, remaining more acidic for a long time. During SD both alkaline and acidic shifts were smaller than in normal Ringer's solution, with the acidic shift being the more affected (Fig. 2).

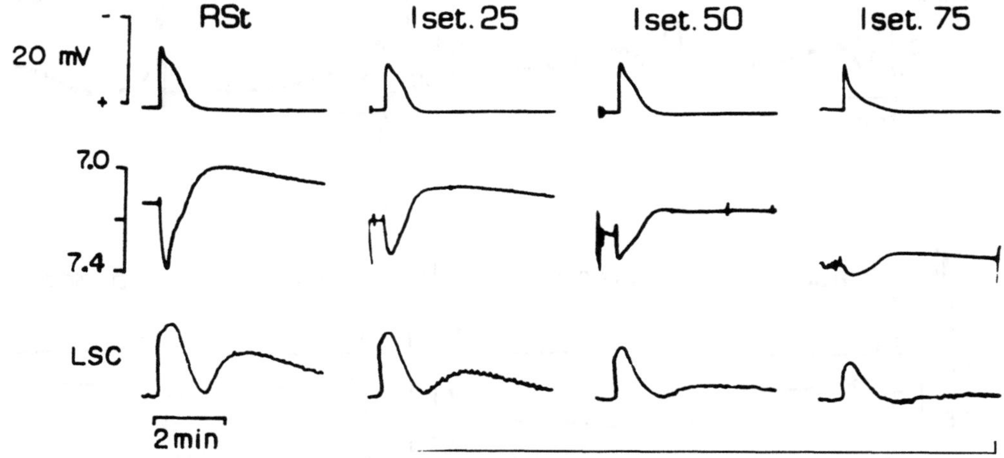

Fig. 2. Effects of Cl^- replacement by isethionate (25%, 50%, and 75%) on steady state pH_o as well as on pH_o changes during SD. ISM location: IPL. Changes in potential and scattered light are also shown

pH_o Changes Under Different Metabolic Conditions

As pointed out previously (Vercesi and Martins-Ferreira 1983), isolated chick retina is more sensitive to glucose concentration than to O_2 in the medium, when an effort is being made to obtain successive and reproducible waves of SD. With the aim of examining the influence of O_2 and glucose utilization by the retina for the maintenance of H^+ homeostasis, a set of experiments was performed using the following metabolic conditions: *substitution of O_2 by N_2, lowering of the glucose concentration, and presence of inhibitors of glycolysis.* In this set of experiments preparations were superfused with the test solutions during 60 min and then with RSt, unless deterioration had occurred. In another set of experiments the effects of temperature were studied.

Substitution of O_2 by N_2. In preparations superfused with Ringer's solution saturated with 100% N_2, the only noticeable difference from control conditions was a small acidification of the resting retina and a decrease of tissue susceptibility to initiation of SD. The time course and amplitude of pH_o changes during SD were similar to those obtained with control solutions (Fig. 3a).

Lowering the Glucose Concentration. When glucose in the Ringer's solution was lowered from 30 to 15 m*M* there was a tissue alkalinization in resting conditions, and an increase of tissue susceptibility to the initiation of SD with optical stimuli. The pH_o changes during SD were not, however, significantly different from those recorded with control solutions. When

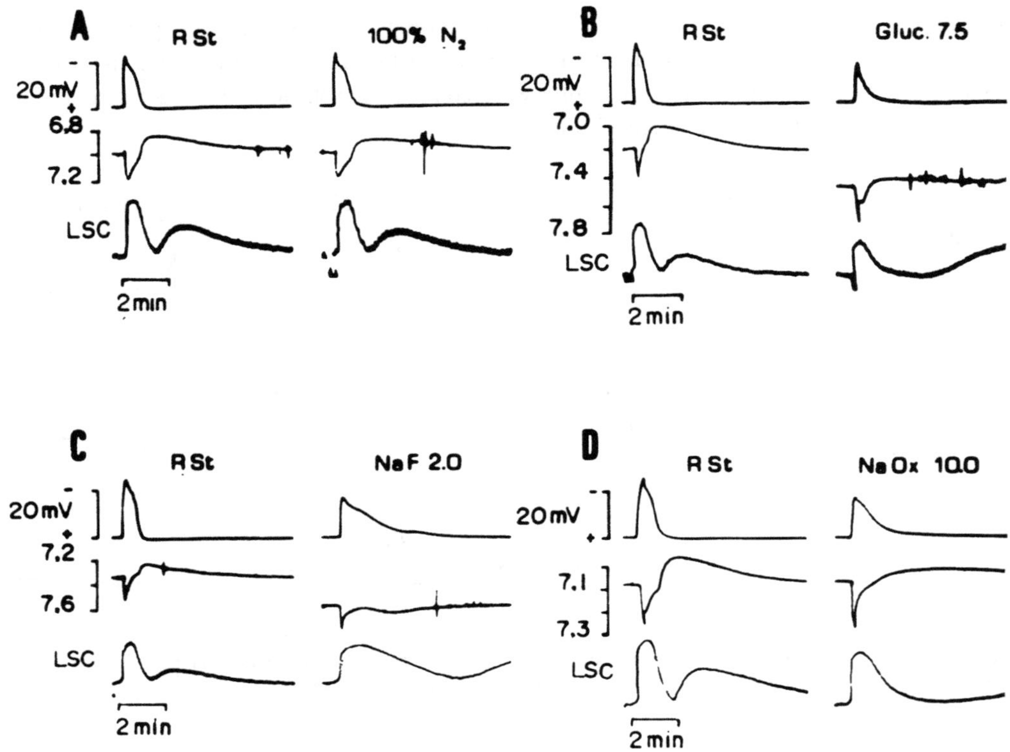

Fig. 3A-D. Effects of different metabolic conditions on steady state pH_o as well as on pH_o changes during SD. ISM location: IPL. **A** Superfusion with Ringer's solution gassed with 100% N_2; **B** lowering of glucose concentration from 30 to 7.5 m*M*; **C** addition of 2 m*M* NaF to RSt; **D** addition of 10 m*M* sodium oxamate to RSt

the retina was superfused with Ringer's solution containing 7.5 m*M* glucose, alkalinization in resting conditions was more conspicuous. During SD, although the amplitude of the alkaline shift was similar to that recorded in control solutions, the acidic shift almost completely disappeared (Fig. 3b). Lowering the Ringer's solution glucose concentration caused a significant instability of the retina, which became more evident the lower the glucose concentration was and usually there were recurrent spontaneous waves of SD. Also, tissue deterioration occurred earlier than in control conditions.

Use of Glycolysis Blockers. Fluoride and oxamate are known to block glycolysis at different levels: fluoride at phosphoenolpyruvate formation, and oxamate at lactate formation (Novoa et al. 1959; Lehninger 1975). To test how this metabolic pathway influences retinal H^+ homeostasis, we examined the effects of these blockers on tissue pH_o under resting conditions and during SD. The addition of 2 m*M* sodium fluoride to Ringer's solution caused a significant increase in baseline pH_o, while sodium oxamate was ineffective. During SD none of these substances affected the alkaline shift amplitude, but they had a huge effect on the acidic shift: sodium fluoride caused its complete disappearance and sodium oxamate a significant decrease (Fig. 3c, d).

Effects of Temperature. In another set of experiments with the aim of testing the participation of metabolic events in retinal H^+ homeostasis in resting conditions and during SD, we examined the effects of lowering the bath temperature. It was previously reported that alterations of bath temperature caused accentuated modifications in susceptibility of the retina to initiation of SD, as well as in velocity of propagation, depth profile of field potentials, and ion translocations during SD (Martins-Ferreira et al. 1974; do Carmo and Martins-Ferreira 1984). The ISM was positioned at depths of 100 μm and 180 μm and the pH_o was measured while changing the bath temperature from 30° to 23°C, at a rate of 2°C/15 min. The decrease in temperature caused a tissue alkalinization in resting steady state conditions in both layers. The effect on pH_o changes during SD was dependent on the ISM depth: when the ISM was placed in the proximal retina, the amplitude of the alkaline shift at 23°C was a little smaller than that recorded at 30°C, and sometimes it split into two peaks; the amplitude of the acidic shift was higher than that recorded at 30°C (Fig. 4a). With the ISM located in the distal retina, both shifts increased in magnitude (Fig. 4b).

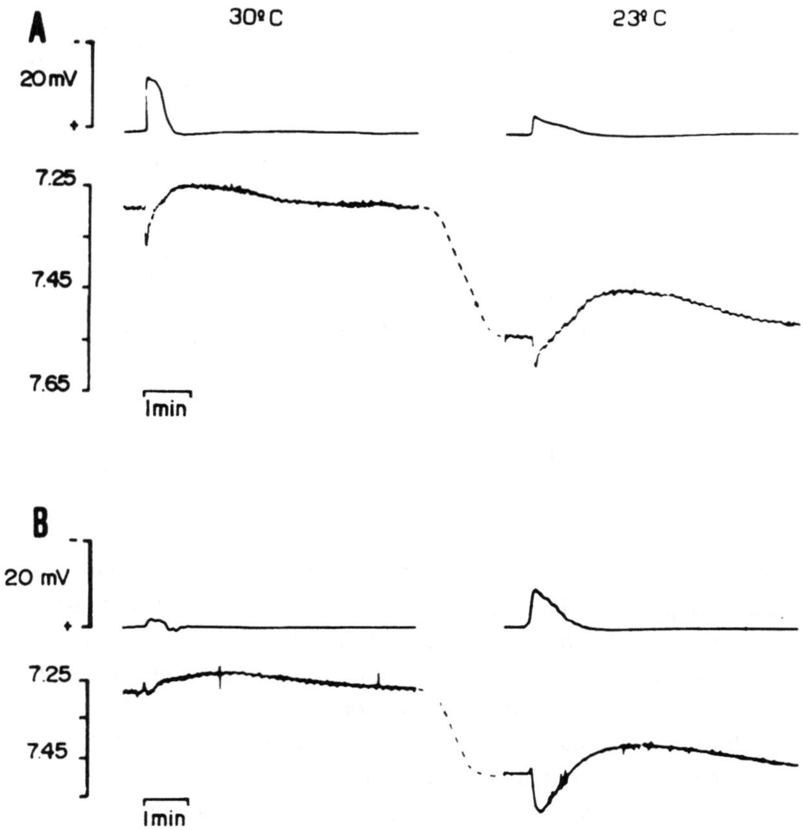

Fig. 4A,B. Effects of temperature on steady state pH_o as well as on pH_o changes during SD. ISM location: **A** proximal retina; **B** distal retina

Effects of Inhibitors

Carbonic Anhydrase Inhibitor. Carbonic anhydrase is found in brain predominantly in the glia and its role in pH_o -buffering power in other structures of the CNS has been reported (Kraig et al. 1983; Mutch and Hansen 1984). Here we examined its role in isolated chick retina pH_o using the inhibitor acetazolamide. Superfusion of preparations with solutions containing up to 8.0 m*M* acetazolamide caused a significant alkalinization of the retina and a remarkable increase in amplitude of both alkaline and acidic shifts during SD (Fig. 5). In a few experiments the treatment with acetazolamide was followed by the appearance of occasional spontaneous waves of SD, but there was no sign of retina deterioration.

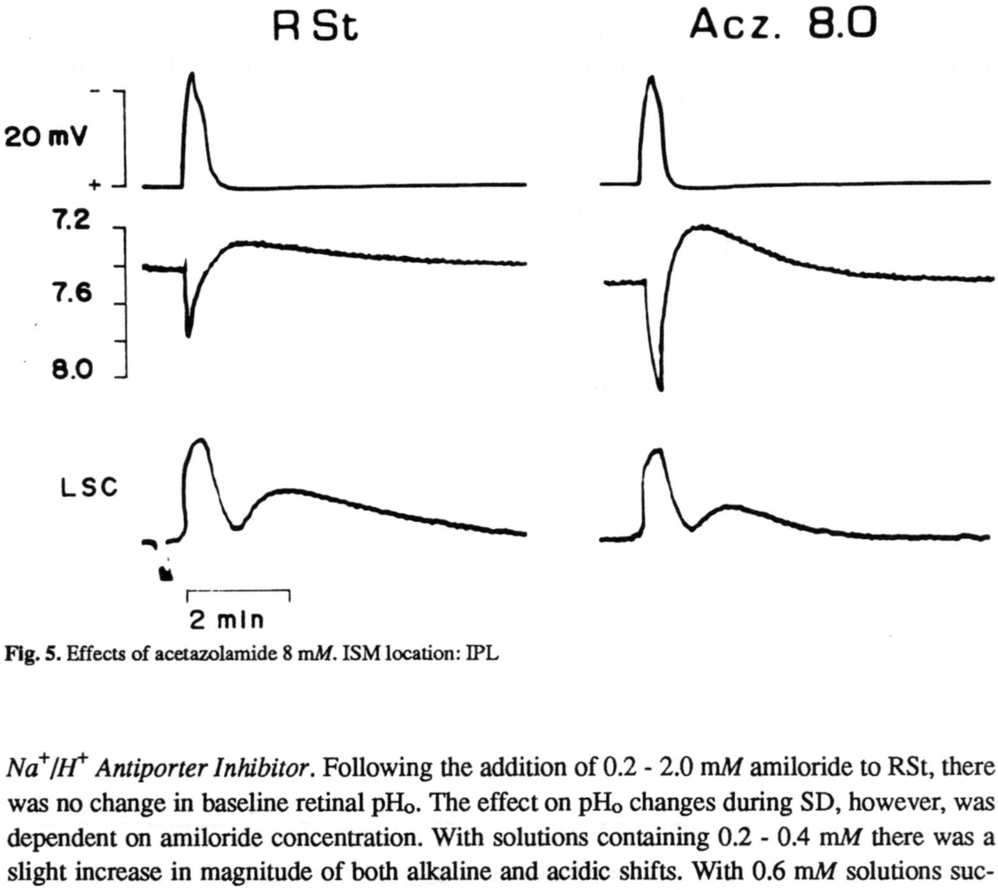

Fig. 5. Effects of acetazolamide 8 m*M*. ISM location: IPL

Na^+/H^+ Antiporter Inhibitor. Following the addition of 0.2 - 2.0 m*M* amiloride to RSt, there was no change in baseline retinal pH_o. The effect on pH_o changes during SD, however, was dependent on amiloride concentration. With solutions containing 0.2 - 0.4 m*M* there was a slight increase in magnitude of both alkaline and acidic shifts. With 0.6 m*M* solutions successive waves of spontaneous SD were observed, and the shifts were smaller than in the control because the intervals between the SD waves were too short (Fig. 6b). With 1.2 m*M* amiloride the retina became less susceptible to initiation of SD and, once it began, the wave propagated only a few millimeters. With 2.0 m*M* amiloride the tissue became refractory and SD could not be initiated even when stimulated mechanically or with 40-m*M* KCl pulses. These effects were reversed 30-40 min after resuperfusion of the retina with RSt.

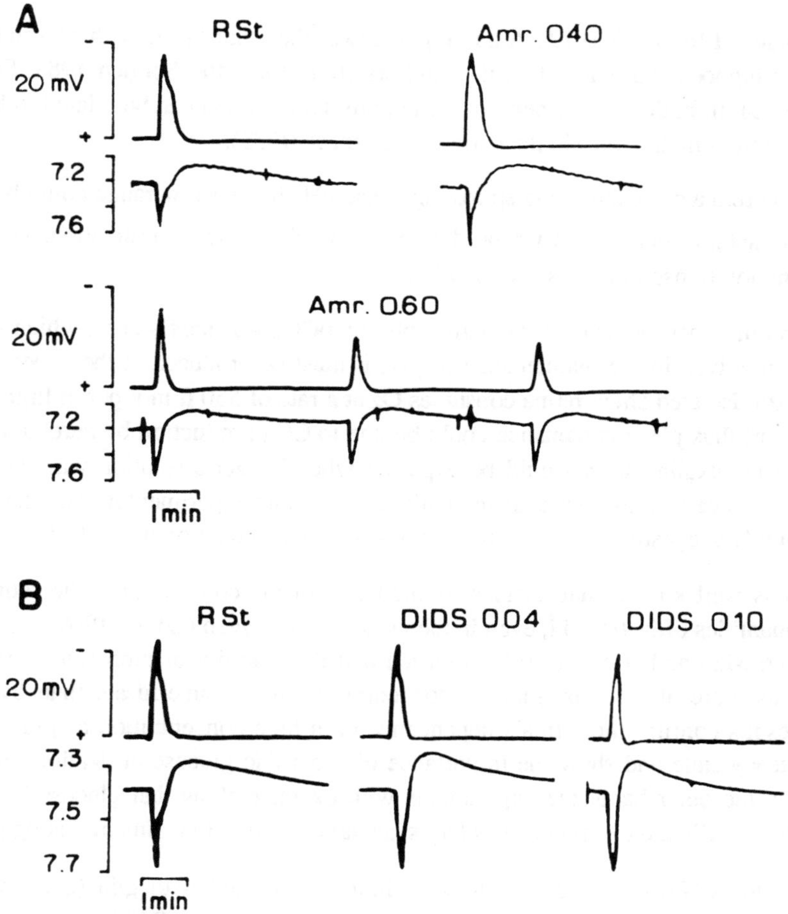

Fig. 6A,B. Effects of **A** amiloride 0.4 and 0.6 m*M* and **B** DIDS 0.04 - 0.1 m*M*. ISM location: IPL

Cl^-/HCO_3^- Antiporter Inhibitor. When 0.02-0.1 m*M* DIDS was added to RSt there was no change in baseline pH_o. During SD the shifts were shorter than in control and there was no modification in the amplitude of the alkaline shift, but the amplitude of the acidic shift was slightly increased. With 0.2 m*M* DIDS there was a retinal alkalinization in resting conditions and the preparation became less susceptible to the triggering of SD, which occasionally could not be evoked even by mechanical stimulation. These effects were partially reversed after 120 min of resuperfusion with RSt, but clear signs of deterioration were observed.

Discussion

One of the most important features of these data was the retina's ability to maintain pH_o at 7.16 at the inner plexiform layer when superfused with a solution of pH about 9.0, and to show an acidification of no more than 0.2 pH units when the pH of Ringer's solution was changed from 9.0 to 7.5. A tissue pH_o lower than that of the bathing solution was reported for hippocampal slices by other authors (Balestrino and Somjen 1988; Schiff and Somjen 1985). In both types of nervous tissue, this acidification can be related to liberation and diffusion of metabolites in the extracellular space (ECS).

The retina's capacity to keep resting tissue pH_o in a narrow range could be due to a number of factors: metabolic CO_2 production, anaerobic glucose utilization, HCO_3^- dehydration, and ion transport across cell membranes.

In our experiments the "standard" solution pCO_2 was nearly zero, which means that if CO_2 is involved in the maintenance of pH_o it must be produced at the expense of cell metabolism. Isolated chick retina consumes O_2 at a rate of 550 μ mol per milligram dry weight per hour; thus pH_o maintenance could be due to CO_2 production by aerobic metabolism, and a tissue alkalinization would be expected when Ringer's solution was saturated with 100% N_2. Since pH_o did not change significantly in this experimental condition, these results are highly suggestive that H_o^+ homeostasis is not dependent on the cells' CO_2 production.

It is well known that besides its high rate of O_2 consumption, the retina produces huge quantities of lactic acid, even in the presence of oxygen (Lolley 1969). Furthermore, Vercesi and Martins-Ferreira (1983) observed that the viability of preparations in SD experiments was more affected by glucose concentration in the superfusate than by lack of O_2. Our results confirm the retina's dependence on a high concentration of glucose to maintain its steady state and show the importance of anaerobic glucose utilization for H^+ homeostasis. On the other hand, the experiments with oxamate show that glucose breakdown does not necessarily have to proceed as far as lactate formation to maintain resting pH_o.

In the CNS, carbonic anhydrase is found associated with glia (Giacobini 1962) and, in retina, with Müller cells (Lolley 1969). Its importance in CNS homeostasis was assessed by utilization of the specific inhibitor acetazolamide. When applied directly on CNS tissue in vivo, acetazolamide causes an extracellular alkalinization and its effects are related to reduction in the average reaction time for the hydration of CO_2 (Kraig et al. 1983; Mutch and Hansen, 1984). Since pCO_2 is low, the equilibrium of the carbonic anhydrase reaction lies in the direction of HCO_3^- dehydration. Although further studies have to be made, Müller cells seem to be actively involved in retina H_o^+ homeostasis through the action of carbonic anhydrase on HCO_3^- dehydration.

pH_o Changes During Spreading Depression

In the isolated retina, pH_o shifts during SD resemble those reported for other brain tissues in vivo and in vitro (Kraig et al. 1983; Mutch and Hansen 1984; Nicholson et al. 1985; Balestrino and Somjen 1988; Rice and Nicholson 1988). In the retina the phenomenon is accompanied by ion translocations and changes in ECS volume (do Carmo and Martins-Ferreira 1984) which could interfere with pH_o values.

pH_o Alkaline Shift. The alkaline shift was rapid and coincident with the negative slow variation of baseline voltage, which is accompanied by marked changes in the extracellular ionic composition, and a significant decrease in extracellular space (Kraig and Nicholson 1978; Phillips and Nicholson 1979; do Carmo and Martins-Ferreira 1984).

By comparing measurements with LIX microelectrodes of ($[K^+] + [Na^+] + 2[Ca^{2+}]) - [Cl^-]$ at resting baseline and during SD, Kraig and Nicholson (1978) found a base excess of 29-30 mEq. In isolated chick retina similar measurements made at IPL revealed a base excess of 30 mEq (do Carmo and Martins-Ferreira 1984). Such an excess could account for the alkaline shift seen during SD. This hypothesis is reinforced by the fact that at the distal retina, where there was an acid excess of 15 mEq (do Carmo and Martins-Ferreira 1984), the alkaline shift had disappeared.

Ionic translocations between compartments could also account for the modifications of pH shifts recorded when the retina was cooled. Measuring extracellular $[K^+]$, $[Cl^-]$, and $[Ca^{2+}]$, do Carmo and Martins-Ferreira (1984) observed that a change in bath temperature from 30° to 23°C markedly diminished the ionic changes in the proximal retina and enhanced those in the distal retina, which could be related to the decrease and enhancement of the alkaline shift in the proximal and distal retina, respectively.

The alkaline shift could also be due to a rise in $[HCO_3^-]_o$. A calculated $[HCO_3^-]_o$ increase of about 30 m*M* was reported by other authors (Mutch and Hansen 1984; Kraig and Cooper 1987) and could result from efflux of intracellular HCO_3^- to accompany efflux of intracellular K^+ in order to maintain electroneutrality in the ECS or, if HCO_3^- is impermeant, from a relative rise in its concentration due to a decrease in size of the extracellular space. This last supposition seems to be more probable, since substitution of Cl^- by isethionate caused a significant modification both in extracellular space changes (do Carmo and Martins-Ferreira 1984) and in the amplitude of the alkaline shift. Further experiments are needed to sustain this hypothesis, since when retina was superfused with Ringer's solution containing 3 m*M* Mn^{2+}, which completely abolished the alkaline shift during stimulation of cerebellar parallel fibers (Chesler and Chan 1988), SD could not be evoked by optical, mechanical, or chemical stimuli. Increase in $[HCO_3^-]_o$ would markedly enhance the dehydration of HCO_3^- to CO_2 mediated by carbonic anhydrase (Giacobini 1962), thus damping the alkaline shift, an assumption supported by the results with acetazolamide. The shift could also be damped by stimulation of Cl^-/HCO_3^- or inhibition of Na^+/H^+ exchangers. Enhancement of the alkaline shift by addition of DIDS or amiloride was recorded in the cerebral cortex in vivo and assumed to be due to inhibition of both processes (Mutch and Hansen 1984). These results were not confirmed in our experiments and, in turtle cerebellum in vitro amiloride caused no changes in alkaline shift during stimulation (Chesler and Chan 1988). These data suggest that, at least for in vitro preparations, neither the Na^+/H^+ nor the Cl^-/HCO_3^- antiporter is involved in the generation of the alkaline shift.

pH_o Acid Shift. In isolated chick retina the amplitude and duration of the acid shift during SD is in the range of values reported by several authors for preparation in vivo and in vitro (Lehmenkuhler et al. 1981; Kraig et al. 1983; Nicholson et al. 1985; Rice and Nicholson 1988), and could be explained by an increase in lactic acid extrusion and/or by activation of carrier-mediated ionic transport. It is well known that during SD there is an increase in glycolysis rate and lactic acid content in brain tissue (Křivánek 1961; Quistorff et al. 1979; Mutch and Hansen 1984). Furthermore studies on changes of oxygen tension and O_2 consumption accompanying the phenomenon indicate that the lactic acid increase is probably related to metabolic demand to reestablish the tissue's ionic steady state (Bureš et al. 1974; Vercesi and Martins-Ferreira 1983). In the chick retina the importance of aerobic glucose breakdown for the generation of the acid shift is contested by data obtained with Ringer's solution saturated with 100% N_2. On the other hand, the effects of glycolytic blockers are similar to those of lowering glucose, i.e., a progressive retinal deterioration and disappearance of the acid shift. These data are indicative that the acidic shift is closely correlated with lactic acid production, and that glycolysis is essential to furnish the necessary energy for the recovery of the tissue as well as for the maintenance of its ability to sustain successive waves of SD.

Summary

Ion-selective microelectrodes were used to measure extracellular pH (pH_o) in isolated chick retina under resting conditions and during spreading depression (SD). When superfused with "standard" Ringer's solution saturated with 100% O_2 at pH 8.9, the preparation was able to maintain resting pH_o at 7.16 in the inner plexiform layer. Resting pH_o was not affected by lack of O_2 or by addition of sodium oxamate, amiloride, or 4,4'- diisothiocyanostilbene-2,2'-disulfonic acid (DIDS). Lowering bath temperature or glucose concentration, substitution of Cl^- by isethionate, and addition of sodium fluoride or acetazolamide caused a significant alkalinization of resting pH_o. These data show the high capacity of the isolated retina for maintaining H^+ homeostasis, mainly at the expense of anaerobic glucose metabolism, and the influence of carbonic anhydrase on the buffering properties of nervous tissue. The pH_o changes accompanying SD were usually biphasic, with an initial rapid alkaline shift followed by a smaller and longer acidic shift. Their amplitudes were dependent on microelectrode depth. A reduction in amplitude of the alkaline shift occurred when bath temperature was lowered or Cl^- was replaced by another ion. Lowering glucose concentration, as well as the addition of sodium fluoride or sodium oxamate affected only the acidic shift, causing its disappearance. Addition of acetazolamide enhanced both shifts; amiloride and DIDS had no effect on either. These shifts resembled those reported in other preparations and were damped by carbonic anhydrase. The alkaline shift is probably related to strong ion translocations and HCO_3^- dehydration, and the acidic shift is dependent on retinal lactic acid production.

Acknowledgements. We are grateful to Prof. W. Simon (Swiss Federal Institute of Technology, Zurich, Switzerland) for kindly providing the H^+-LIX; to Dr. R. P. Kraig (Cornell

University, NY, United States) for providing some of the drugs and for valuable discussions of the experiments, and to Dr. C. Nicholson (New York University Medical Center, NY, United States) for his helpful comments on the manuscript.

References

Ammann D, Lanter F, Steiner RA, Schulthess P, Shijo Y, Simon W (1981) Neutral carrier based hydrogen ion selective microelectrode for extra- and intracellular studies. Anal Chem 53: 2267-2269

Balestrino M, Somjen GG (1988) Concentration of carbon dioxide, interstitial pH and synaptic transmission on the hippocampal formation of the rat. J Physiol (Lond) 396: 247-266

Bureš J, Burešová O, Křivánek J (1974) The mechanism and applications of Leão's spreading depression of electroencephalographic activity. Academic, New York

Chesler M, Chan CY (1988) Stimulus-induced extracellular pH transients in the in vitro turtle cerebellum. Neuroscience 27: 941-948

do Carmo RJ, Ferreira-Filho CR, Martins-Ferreira H (1985) Liquid unstirred layers and spreading depression (SD) in isolated retina. An Acad Bras Ciênc 57: 528-529

do Carmo RJ, Martins-Ferreira H (1984) Spreading depression of Leão probed with ion-selective microelectrodes. An Acad Bras Ciênc 56: 401-421

Ferreira-Filho CR, Martins-Ferreira H (1983) pH changes in extracellular milieu during retinal spreading depression. Braz J Med Biol Res 16: 458

Ferreira-Filho CR, Martins-Ferreira H (1985) Influence of O_2, glucose and glycolytic blockers on retina H^+ homeostasis and on pH changes during spreading depression. Braz J Med Biol Res 18: 629A

Giacobini E (1962) A cytochemical study of the localization of carbonic anhydrase in the nervous system. J Neurochem 9: 169-177

Kraig RP, Cooper AJL (1987) Bicarbonate and ammonia changes in brain during spreading depression. Can J Physiol Pharmacol 65: 1099-1104

Kraig RP, Nicholson C (1978) Extracellular ionic variations during spreading depression. Neuroscience 3: 1045-1059

Kraig RP, Ferreira-Filho CR, Nicholson C (1983) Alkaline and acid transients in cerebellar microenvironment. J Neurophysiol 49: 831-849

Křivánek J (1961) Some metabolic changes accompanying Leão's spreading cortical depression in the rat. J Neurochem 6: 183-189

Lehmenkuhler A, Zidek W, Staschen M, Caspers H (1981) Cortical pH and pCa in relation to DC potential shifts during spreading depression and asphyxiation. In: Syková E, Hnìk P, Vyklický L (eds) Ion-selective microelectrodes and their use in excitable tissues. Plenum, New York, 225-229

Lehninger AL (1975) Biochemistry, 2nd edn. Worth, New York

Lolley RN (1969) Metabolical and anatomical specialization within the retina. In: Lajtha A (ed) Handbook of neurochemistry, vol 2. 473-504

Lux HD (1974) Fast recording ion specific microelectrodes: their use in pharmacological studies in the CNS. Neuropharmacology 13: 509-517

Martins-Ferreira H (1962) Depressão alastrante na retina. An Acad Bras Ciênc 34: XLIV

Martins-Ferreira H (1983) Spreading depression in chicken retina. In: Ookawa T (ed) The brain and behavior of the fowl. Japan Scientific Society Press, Tokyo, pp 317-333

Martins-Ferreira H, do Carmo RJ (1987) Retinal spreading depression and the extracellular milieu. Can J Physiol Pharmacol 65: 1092-1098.

Martins-Ferreira H, de Oliveira-Castro G, Struchiner CJ, Rodrigues PS (1974) Circling spreading depression in isolated chick retina. J Neurophysiol 37: 773-784

Mutch WAC, Hansen AJ (1984) Extracellular pH changes during spreading depression and cerebral ischemia: mechanisms of brain pH regulation. J Cereb Blood Flow Metab 4: 17-27

Nicholson C, Kraig RP, Ferreira-Filho CR, Thompson P (1985) Hydrogen ion variations and their interpretations in the microenvironment of the vertebrate brain. In: Kessler M, Harrison DK, Höper J (eds) Ion measurements in physiology and medicine. Springer, Berlin Heidelberg New York, pp 229-235

Novoa WB, Winer AD, Glaid AJ, Schwert GW (1959) Lactic dehydrogenase. V - Inhibition by oxamate and by oxalate. J Biol Chem 234: 1143-1148

Phillips JM, Nicholson C (1979) Anion permeability in spreading depression investigated with ion-sensitive microelectrodes. Brain Res 173: 567-571

Quistorff B, Gjedde A, Hansen AJ (1979) Spatial analysis of the freeze-trapped brain provides for temporal evolution of an event: metabolic, electrical and blood flow changes during spreading depression. Acta Physiol Scand 105: 42A

Rice ME, Nicholson C (1988) Behavior of extracellular K^+ and pH in skate (*Raja erinacea*) cerebellum. Brain Res 461: 328-334

Schiff SJ, Somjen GG (1985) Overshoot of oxygen pressure in post hypoxic brain tissue: a re-evaluation. Brain Res 344: 150-153

Vercesi A, Martins-Ferreira H (1983) Oxygen and glucose requirements in chick retina spreading depression. An Acad Bras Ciênc 55: 309-316

Detection of Magnetic Fields from Isolated Avian Retina During Spreading Depression

Y. C. Okada[1], R. J. do Carmo[2], H. Martins-Ferreira[2], and C. Nicholson[3]

[1] Departments of Neurology and Physiology, University of New Mexico School of Medicine and MEG Center, V A Medical Center, 2100 Ridgecrest Drive, Albuquerque, NM 87108, USA.
[2] Department of Neurobiology, Institute of Biophysics, Federal University of Rio de Janeiro, Ilha do Fundão, 21941 Rio de Janeiro, Brazil.
[3] Department of Physiology and Biophysics, New York University Medical Center, 550 First Ave., New York, NY 10016 USA

Introduction

Spreading depression (SD) of Leão (1944) has been implicated as the underlying process responsible for the aura that is experienced in patients suffering from classic migraine (Lauritzen 1987). Recent evidence comes primarily from the study of regional cerebral blood flow (Olesen et al. 1981; Lauritzen et al. 1983; Lauritzen and Olesen 1984). Electroencephalographic (EEG) and magnetoencephalographic (MEG) studies have been carried out to obtain direct electrophysiological signs of SD during the aura (Shoenen et al. 1987; Tepley et al. 1989). These studies have reported reduction of spontaneous EEG and MEG activity during such an episode. A DC-coupled MEG study has also been carried out on migraine patients (Tepley et al. 1989). The EEG and MEG results are, however, difficult to interpret in terms of SD because of insufficient understanding of the currents that can give rise to such signals during an SD.

Toward Understanding the Currents Underlying MEG Signals of SD

SD can produce a magnetic field that is sufficiently strong to be detectable without averaging at a distance of as much as several centimeters from the tissue (Okada et al. 1987, 1988). In these initial studies, the DC-coupled magnetic field was measured above the surface of a bath containing an isolated turtle cerebellum. The spatial distribution of the magnetic field over the bath surface indicated that the signal was due to currents perpendicular to the cerebellar surface, implying that the underlying currents were oriented in the longitudinal direction of the major principal core conductors of the cerebellum such as the Purkinje cells, ascending axons of the granule cells and ependymal astrocytes. It was unclear, however, from these studies how the magnetic field was related to the currents produced by SD, since the site of initiation of SD and its spatial pattern of propagation were not determined.

On a priori grounds two types of currents may contribute to the external magnetic field during SD (Okada 1990). Consider that an SD is initiated in some region of a neuronal tissue and propagates away from this site at a speed of 2 mm/min, as is commonly the case in many preparations (Bureš et al. 1974). The SDs in in vivo and in vitro preparations last about 3 min (Nicholson 1984). The tissue of several millimeters in width undergoing an episode of SD characteristically exhibits a large potential gradient along the longitudinal direction of principal core conductors such as pyramidal cells in the cortex (Nicholson and Kraig 1981; do Carmo and Martins-Ferreira 1984). Thus, one should expect currents perpendicular to the surface of such a tissue. The results from the turtle cerebellum indicate that indeed such currents can produce a magnetic field.

The magnetic field may also be due to a second class of currents oriented tangentially to the tissue surface. Across the wave front of SD there is a large extracellular potential (φ) drop (Leão 1947) of as much as –30 to –40 mV (Bureš et al. 1974). The extracellular conductivity σ is also reduced by a factor of about 10%-300% (Leão and Martins-Ferreira 1953, 42%; Van Harreveld and Ochs 1957, 17%; Freygang and Landau 1955, 10%-20%; Rank 1964, 60%; Hoffman and Clark 1973, 200%-300%; Ferreira-Filho and Martins-Ferreira 1982, 30%-40%). Thus, the wave front can serve as a conductivity boundary with a potential drop across it. Such a boundary is known to serve as a generator of magnetic field (Geselowitz 1970; Plonsey 1981). The magnetic field B in such a case should be proportional to $\Delta \Phi \Delta \sigma$ for a tissue with a uniform value of Φ and σ across the cross-sectional area of the tissue. This equivalent current generator of the magnetic field should be located at the wave front and directed opposite to the direction of propagation (Okada 1990). It is, therefore, necessary to asses the relative contribution of this second class of currents to the external magnetic field in order to interpret the magnetic field of SD from an in vivo preparation.

In Vitro Avian Retina Preparation

The isolated chick retina preparation is ideally suited for analyzing the above issues since it is possible visually to observe the pattern of propagation of SD by the change in color of the tissue due to alterations in light-scattering during SD (Martins-Ferreira and Oliveira-Castro 1966; Martins-Ferreira et al. 1974). As in the case of turtle cerebellum, it is easy to prepare a flat, rectangular piece of retina, attached either to the sclera or to a black filter paper. The preparation can then be suspended in a bath containing a physiological saline just as for the cerebellum. Cellular processes in the retina are oriented along the direction perpendicular to the retina in one case and along the direction parallel to the retina in the other case. Thus, spatial orientations of the possible generators of the magnetic field are simplified, making it experimentally feasible to interpret the magnetic field in terms of its underlying currents.

Methods

The isolated retina was obtained from 2-month-old chickens (white leghorns), following the procedure of do Carmo and Martins-Ferreira (1984). The eyeballs were enucleated after decapitation, then a sharp razor blade was used to cut away a rectangular piece of retina (about $0.25 \times 5 \times 15$ mm). The retina was either attached to the sclera, or removed from the sclera and adhered to a rectangular black tissue paper of nearly the same size. The tissue was then suspended vertically or horizontally by a nylon net in a large plastic chamber (Fig. 1) (Okada and Nicholson 1988). The chamber was filled with physiological saline consisting of: NaCl, 100 m*M*; KCl, 6 m*M*; NaH_2PO_4, 1 m*M*; $NaHCO_3$, 30 m*M*; $CaCl_2$, 1 m*M*; $MgCl_2$, 1 m*M*; and glucose, 30 m*M*. SDs were initiated in some cases mechanically, pressing against the retina with a sharp glass micropipette. In initial experiments SDs were also triggered with a local application of KCl or glutamate. The occurrence of SD was visually monitored by the experimenter.

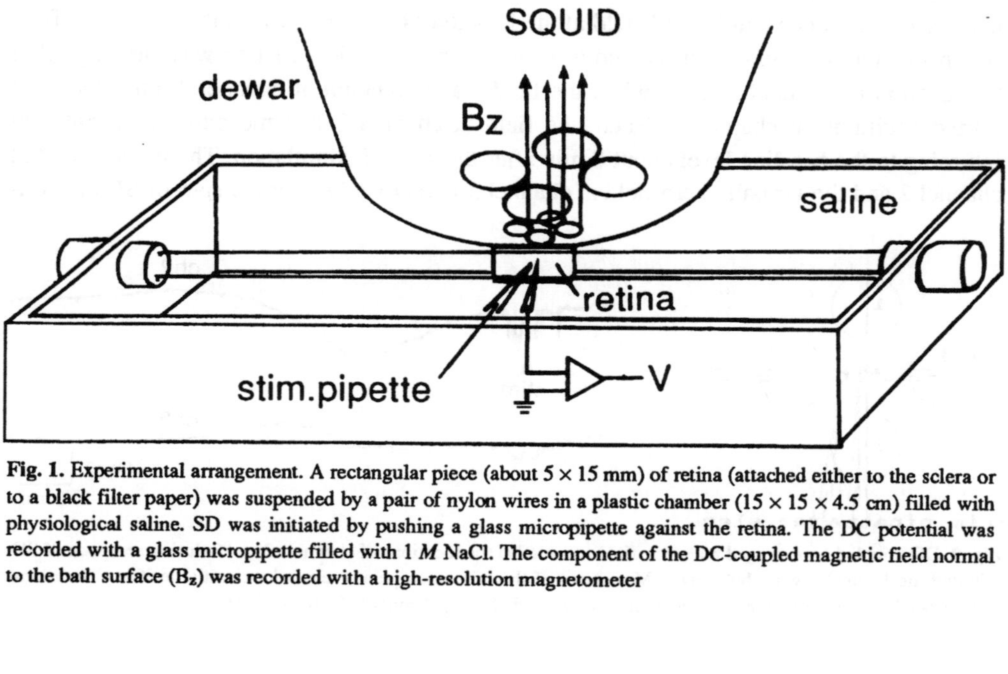

Fig. 1. Experimental arrangement. A rectangular piece (about 5 × 15 mm) of retina (attached either to the sclera or to a black filter paper) was suspended by a pair of nylon wires in a plastic chamber (15 × 15 × 4.5 cm) filled with physiological saline. SD was initiated by pushing a glass micropipette against the retina. The DC potential was recorded with a glass micropipette filled with 1 *M* NaCl. The component of the DC-coupled magnetic field normal to the bath surface (B_z) was recorded with a high-resolution magnetometer

The magnetic field measurements were carried out with a four-channel, high-resolution magnetic field detector specifically developed for measurements of magnetic fields close to the source (Buchanan et al. 1989). Our system consisted of four axial first-order asymmetric gradiometers coupled to DC-SQUIDs (superconducting quantum interference devices). The pickup coil diameter was 4.0 mm, the baseline 16 mm, and the channel separation 6.0 mm. The minimum distance between the pickup coil and the outer surface of the cryostat was 1.1 mm. The noise level of the channels was 0.050 $pT/\sqrt{Hz}$. The DC-coupled electrical potential was also monitored conventionally in some experiments with a glass micropipette filled with 1 *M* NaCl. The magnetic field and electrical potential were monitored on a two-channel strip chart recorder and simultaneously on a four-channel digital strip chart recorder. All the experiments were carried out in a magnetically shielded, two-shell, hypermalloy room (inner room, 1.22 × 1.52 × 1.83 m; outer room, 1.83 × 2.29 × 2.44 m; wall thickness, 1.5 mm) located in the Center for Magnetoencephalography.

Results

The basic results are shown in Figs. 2 and 3. Figure 2 illustrates the arrangement for this particular measurement. The retina was held vertically along the axis of channels 2 and with the receptor side facing channel 1. The tissue was slightly off to the right of axis of channels 1 and 3 with the stimulating glass micropipette slightly to the left of the center of the tissue. The pipette was pushed against the tissue with a micromanipulator until the experimenter saw an SD in the form of a propagating circular white patch. The unaveraged magnetic fields recorded from channels 2 and 3 during this epoch are shown in Fig. 2. The initial deflections are artifacts due to the movement of the stimulator. As the SD propagated over the tissue, it produced an inward field at the location of channel 2, with the maximum field occurring at about 2 min after the end of stimulation. The field variation was virtually nil at the location of channel 3, even when the field was maximum at channel 2, until the field started declining at channel 2. Then, the field at channel 3 became outward, eventually returning to the baseline level about 8 min after the start of stimulation. The inward field at channel 2 and the virtually zero field at channel 3 immediately after the start of SD are con-

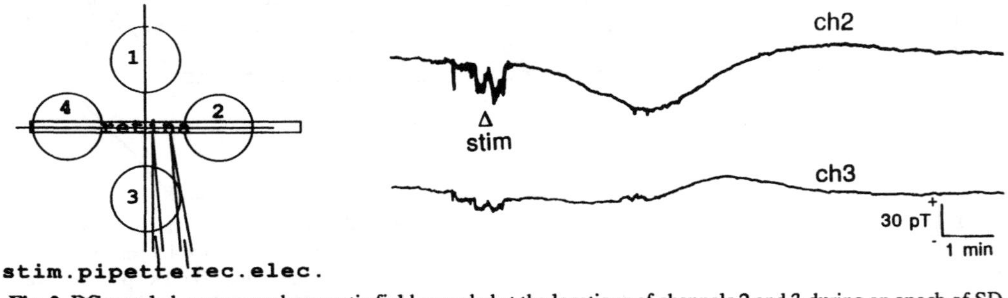

Fig. 2. DC-coupled, unaveraged magnetic field recorded at the locations of channels 2 and 3 during an epoch of SD initiated mechanically with the pipette. Movement of the micromanipulator caused the initial artifact. The retina was suspended vertically in the bath with the receptor side facing channel 1. Outward field = +

sistent with the fact that the site of SD initiation was centered over channel 3, but was to the left of channel 2. The outward field at location 3 can be interpreted as due to the SD in progress on the right side of the retina, since SD reached the left edge of the retina before reaching its right edge.

The maximum field amplitude was 30 pT in this record and as much as 50 pT in other records (see Fig. 3) at a vertical distance of 4 mm for the component of the magnetic field normal to the bath surface. This magnitude is consistent with the results obtained in initial experiments conducted with a single-channel, second-order magnetometer. Due to differences in the design of the magnetometer, the minimum measurement distance was 18 mm in these initial experiments. In this case the field amplitude was 1.8 pT at a distance of 18 mm. Assuming that the fields were due to a dipolar current generator whose field amplitude falls as an inverse of the square of the distance, the field amplitude should be 36 pT at a distance of 4 mm for such a field. The temporal shape and duration of the field were comparable in both sets of experiments. SDs triggered by local applications of 1 *M* KCl and glutamate also produced magnetic fields in these initial experiments that were comparable in shape and magnitude to those produced by the mechanical stimulation.

The magnetic field reversed its polarity, being directed outward at locations 2 and 3, when the tissue was moved 9 mm farther to the right as shown in Fig. 3. This field polarity indicates that the underlying current generator was directed from the side of the inner- to the outer-plexiform layer, since such a generator should produce an inward field to its right and an outward field to its left on the bath surface. The field was weaker at location 3 than at location 2, as expected, since the field extremum should be near location 2 for this measurement distance of 4 mm. The DC potential shown in Fig. 3 provides evidence, in addition to the light-scattering change of the tissue, that an SD was indeed generated in this condition.

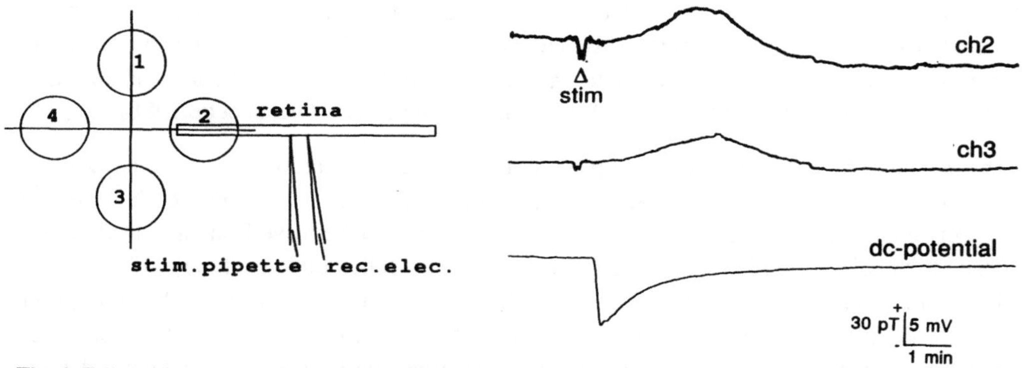

Fig. 3. DC-coupled, unaveraged magnetic field recorded simultaneously with the DC potential. The retina was moved 9 mm to the right from the position shown in Fig. 2. Note polarity reversal at channel 2

No reliable change in the magnetic field was recorded when the retina was rotated along the axis of channels 2 and 4 so that the tissue was suspended horizontally, about 2 mm from the detecting coils. In such an experiment the SD was initiated from one side of the retina, mechanically, and its initiation was visually confirmed. The stimulation initially produced a circular pattern of SD near one edge of the tissue. Then, it propagated along one direction with an arch-shaped wave front. We expected that the equivalent source was oriented perpendicular to the wave front and directed opposite to the direction of propagation. Thus, the magnetic field was expected at locations 1 and 3. Despite our effort to optimize the measurement location, there was no reliable magnetic signature of SD in this condition, above a baseline noise level of about 1 pT.

Discussion

We found a strong magnetic field when the retina was oriented vertically. The polarity reversal observed in the magnetic field on two sides of the retina showed that the field was due to currents directed perpendicularly to the retinal surface during SD. Apparently the transretinal current is strong during SDs, just as for the transcortical current that was found to be strong in the isolated turtle cerebellum during SDs (Okada et al. 1987, 1988).

Furthermore, the polarity of the field was such that the underlying current generator was directed from the side of inner plexiform layer to the outer plexiform layer. This inferred direction of the underlying current is consistent with the laminar profile of the extracellular potential found in the chick retina during SD (do Carmo and Martins-Ferreira 1984). In the case of avian retina, the location of the strongest current sink was inferred to be in the inner plexiform layer. This result is again consistent with the direction of the underlying transcortical current generator inferred from the magnetic field (Okada et al. 1988) and from laminar potential profile (Lauritzen et al. 1988) during SD in the turtle cerebellum.

The magnitude of the field normal to the bath surface was as much as 50 pT at a vertical distance of 4 mm. At a measurement distance of 18 mm, the field amplitude was about 1.8 pT. The results are consistent with the conclusion that there is a strong dipolar term in the transretinal current generator during SD. The magnitude was, however, weaker for the retina than for the turtle cerebellum; the field was as much as 14 pT at a vertical measuring distance of 16 mm for the field produced by a transcortical current (Okada et al. 1988).

The magnetic field was very small, if present, in the case of the retina suspended horizontally. Several reasons can be mentioned for this result. First, our theory could be incorrect in expecting an equivalent current generator oriented normal to the wave front and directed opposite to the direction of propagation. Or, we might have failed to detect the field because of the choice of measurement locations, even if the field was strong enough. We have made an effort without success, however, to measure magnetic fields from several key locations over the retina. Alternatively, it can also be taken as a positive confirmation of our theory. We expected that the field due to the equivalent source at the wave front was weak based on the following considerations. The net current dipole moment of the equivalent dipole current generator was calculated to be: $\Delta \Phi \Delta \sigma A = 20$ mV $\times$ 0.05 S/m $\times$ 0.2 mm $\times$ 5 mm = 0.1 nAm, where A is the cross-sectional area of the retina across the wave front. The value for $\Delta \sigma$ assumes that the conductivity decreases by 40% from a baseline level, where the baseline was taken to be 0.13 S/m (Ferreira-Filho and Martins-Ferreira 1982). This dipole can give rise to a magnetic field of 1 pT at a measurement distance of 2 mm, which is close to the baseline variation of the magnetic field. We are in the process of reducing the DC noise of the instrument output in order to obtain a positive confirmation of our theory.

The above results indicate that the magnetic field due to transcortical currents may be stronger than the field due to the secondary source at the wave front of a propagating SD tangential to the cortical surface. The magnetic field has been successfully measured from an in vivo rabbit with a magnitude of about 2 pT at a measurement distance of about 2 cm (Gardner-Medwin et al. 1989); however, it is unclear from the published records whether the field was due to a transcortical current or not.

Our study shows that a simple preparation such as the avian retina appears to be useful in producing the basic understanding of the currents that can give rise to the magnetic field during SD. This is due not only to the simplicity of the geometry of the preparation, but also to the fact that the spatial propagation of SD can be visually monitored with a naked eye. Understanding of the origin of magnetic field during SD in the preparation should be useful in interpreting more complex field patterns from in vivo or intact preparations during SD or during a migraine aura.

Acknowledgement. We thank the generous support from NINDS under grant NS21149, National Science Foundation under grant DIR-8820556, Research Advisory Group of Department of Veterans Affairs, and the CNPq - Blue Ribbon grant. We also thank Dr. Jin-chu Huang for his assistance during the construction phase of the shielded room needed for this study.

Summary

The magnetic field associated with spreading depression (SD) was studied in the isolated avian retina, using a newly developed high-resolution magnetic field detector. The field was detected over the surface of a bath containing a flat, rectangular piece (about 5 × 15 mm) of retina from a 2-month-old chicken (white leghorn). When the retina was suspended vertically, a strong DC-coupled magnetic field (about 50 pT along the direction normal to bath surface) was observed during an SD. The magnetic field was barely detectable, however, when the retina was placed horizontally in the bath, just below the detector. The results can be interpreted in terms of two types of underlying currents. The transretinal current perpendicular to the retinal surface can explain the strong magnetic field observed when the retina was suspended vertically. The transverse current at the wave front of SD may produce an external magnetic field when the retina is suspended horizontally, but the result indicates that this field was weak. These results have implications for the interpretation of magnetic signals from the brain in general: (1) the DC-coupled magnetic field can be strong and, therefore, useful in detecting electrophysiological signs of SD during classic migraine aura if SD is indeed responsible for the aura; and (2) the magnetic field during SD may primarily reflect transcortical currents in the cortex.

References

Buchanan DS, Crum DB, Cox D, Wikswo JP (1989) MicroSQUID: a close-spaced four channel magnetometer. In: Williamson SJ, Hoke M, Stroink G, M Kotani (eds) Biomagnetism, Plenum, New York, pp 677-679

Bureš J, Burešová O, Křivánek J (1974) The mechanism and applications of Leão's spreading depression of electroencephalographic activity. Academic, Prague

do Carmo RJ, Martins-Ferreira H (1984) Spreading depression of Leão probed with ion-selective microelectrodes in isolated chick retina. Ann Acad brasil Cienc, 56: 401-421

Ferreira-Filho CR, Martins-Ferreira H (1982) Electrical impedance of isolated retina and its changes during spreading depression. Neuroscience, 7: 3231-3239

Freygang WH, Landau WM (1955) Some relations between resistivity and electrical activity in cerebral cortex of cat. J Cell Comp Physiol, 45: 377-392

Gardner-Medwin AR, Tepley N, Barkley GL, Moran J, Nagel-Leiby S, Simkins RT, Welch KMA (1989) Observation of spreading depression in rabbits. In: Williamson SJ, Hoke M, Stroink G, Kotani M, (eds) Biomagnetism. Plenum, New York, pp 323-326.

Geselowitz DB (1970) On the magnetic field generated outside an inhomogeneous volume conductor by internal current sources. IEEE Trans Magn 6: 346-347

Hoffman CJ, Clark FJ (1973) Intracortical impedance changes during spreading depression. J Neurobiol 4: 471-486

Lauritzen M (1987) Cortical spreading depression as a putative migraine mechanism. Trend Neurosci 10: 8-13

Lauritzen M, Olesen J (1984) Regional cerebral blood flow during migraine attacks by xenon-133 inhalation and emission tomography. Brain 104: 447-467

Lauritzen M, Olesen TS, Lassen NA, Paulson AB (1983) Changes in regional cerebral blood flow during the course of classic migraine attacks. Ann Neurol 13: 633-641.

Lauritzen M, Rice ME, Okada YC, Nicholson C (1988) Quisqualate, kainate and NMDA can initiate spreading depression in the turtle cerebellum. Brain Res 475: 317-327

Leão AAP (1944) Spreading depression of activity in the cerebral cortex. J Neurophysiol 7: 359-390

Leão AAP (1947) Further observation on the spreading depression of activity in cerebral cortex. J Neurophysiol 10: 409-419

Leão AAP, Martins-Ferreira H (1953) Alteração da impedância elétrica no decurso da depressão alastrante da atividade do cortex cerebral. An Acad Bras Ciênc 25: 259-266

Martins-Ferreira H, Oliveira Castro G (1966) Light scattering changes accompanying spreading depression in isolated retina. J Neurophysiol 29: 715-726

Martins-Ferreira H, Oliveira Castro G, Struchiner CJ, Rodrigues PS (1974) Circulating spreading depression in isolated chick retina. J Neurophysiol 29: 715-726

Nicholson C (1984) Comparative neurophysiology of spreading depression in the cerebellum. Ann Acad Brasil Ciênc 56: 481-494

Nicholson C, Kraig RP (1981) The behavior of extracellular ions during spreading depression. In: Zeuthen T (ed) The Application of Ion-Selective Microelectrodes, Elsevier, New York; 217-238

Okada YC (1990) Magnetoencephalography as a non-invasive tool for electrophysiological characterization of auras in classic migraine. In: Sato S (ed) Advances in neurology: Magnetoencephalography, vol 54. Sato S(ed) Raven, New York, pp 133-140

Okada YC, Nicholson C (1988) Magnetic evoked field associated with transcortical currents in turtle cerebellum. Biophys J 53: 723-731

Okada YC, Lauritzen M, Nicholson C (1987) Source models and physiology. Phys Med Biol 32: 43-51

Okada YC, Lauritzen M, Nicholson C (1988) Magnetic field associated with spreading depression: a model for the detection of migraine. Brain Res 442: 185-190

Olesen J, Larsen B, Lauritzen M (1981) Focal hyperemia followed by spreading oligemia and impaired activation of rCBF in classic migraine. Ann Neurol 9: 344-352

Plonsey R (1981) Generation of magnetic fields by the human body (theory). In: Erne SN, Hahlbohm HD, Lubbig H, (ed) Biomagnetism. Walter de Gruyter. Berlin; pp 177-205

Rank JB Jr (1964) Specific impedance of cerebral cortex during spreading depression, and an analysis of neuronal, neurological and interstitial contributions. Exp Neurol 9: 1-16

Shoenen J, Jamart B, Delwaide PJ (1987) Cartographie electroencephalographique dans les migraines en periodes critique et intercritique. Rev Electroencephalogr Neurophysiol Clin 17: 289-299

Tepley N, Barkley GL, Moran J, Simkins RT, Welch KMA (1989) Observation of spreading cortical depression in migraine patients. In: Williamson SJ, Hoke M, Stroink G, Kotani M, (eds) Biomagnetism Plenum, New York, pp 372-330

Van Harreveld A, Ochs S (1957) Electrical and vascular concomitants of spreading depression. Am J Physiol 189: 159-166

Wikswo JP, Friedman RN, Kilroy AW, van Egeraat JM, Buchanan DS (1989) Preliminary measurements with microSQUID. In: Williamson SJ, Hoke M, Stroink G, Kotani M (eds) Biomagnetism Plenum, New York, pp 681-684

Thalamic and Hippocampal Spreading Depression: Electrophysiological and Behavioral Evidence

V. I. Koroleva, N. A. Gorelova, and L. V. Vinogradova

USSR Academy of Sciences, Institute of Higher Nervous Activity and Neurophysiology, 5a Butlerov St., Moscow 117865, USSR

Introduction

The concept of a possible functional significance of Leão's spreading depression (Leão 1944, 1951) is largely determined by the threshold of this reaction to different kinds of irritation. In normal brain, spreading depression (SD) is considered to be caused only by excessive electrical stimulation or by strong physical or chemical stimuli. At the same time many experimental stimuli used to cause SD are known to include a component which prevents the development of this reaction. Prolonged electrical stimulation (ES), for example, results in blocking of SD (Koroleva and Bureš, 1980; Koroleva and Gorelova 1983). When epileptic drugs are applied locally, the SD threshold increases rapidly with the development of the seizure activity (Purpura 1963; Aquino-Cias et al. 1971; Bureš et al. 1975; Koroleva et al. 1985). Even the chemical agent KCl, widely used for causing SD, contains an ion of chloride, which weakens the depolarizing effect of potassium (Leão 1963; Bureš et al. 1974; Do Carmo and Martins-Ferreira 1984; Guedes 1984). This contradiction is most vivid when the correlation between SD and seizure afterdischarge is considered. On one hand, an afterdischarge can provoke SD in different brain structures (Leão 1944; Futamachi et al. 1974; De Luca et al. 1977; Machek et al. 1977; Heinemann et al. 1977; Bureš and Burešova 1981; Guedes et al. 1987; Stringer and Lothman 1989); on the other hand, it activates the mechanisms which hamper SD development until it is completely blocked (Whielden and Van Harreveld 1950; Van Harreveld and Stamm 1953; Koroleva and Bureš 1982; Koroleva and Gorelova 1983).

In a previous report we showed (Koroleva 1989; Koroleva and Vinogradova 1990) that increasing the frequency and reducing the duration of rhythmical cortical ES provides a way to differentiate between an excitatory and inhibitory effect on SD occurrence.

The aim of this paper was to investigate SD occurrence in subcortical structures as a preseizure response to a weak ES in freely moving animals and the influence of isolated SD in the thalamus and hippocampus on their behavior.

Methods

Animals and Surgery

A total of 70 male white rats, weighing 300-350 g at the time of surgery, were used as subjects. They were housed in groups of five under a normal 12 h light/12 h dark schedule with free access to food and water. All experiments were carried during the light phase. A first series of experiments were carried out on 50 rats, narcotized by pentobarbital (40 mg/kg) or immobilized by *d*-tubocurarine (0.5-1 mg/kg) and locally anesthetized by 2% novocaine.

To carry out experiments on freely moving animals, 20 rats under pentobarbital anesthesia were implanted with electrodes for ES and recording of brain electrical activity. Electrodes were introduced stereotaxically according to the atlas of Fifková and Maršala (1967) using the following coordinates: hippocampus, + 2.5 to + 3.0 mm/2.5 to 3.0 mm/3.2 to 4.0 mm; thalamus + 2.0 to + 3.0mm/2.5 to 3.0 mm/5.0 to 6.0 mm. The animals were allowed at least 7 days to recover from surgery.

Field Potential Recording

Two types of electrodes were used to register the electrical activity of the brain: glass electrodes (tip diameter, 40-50 μm) filled with 0.9% NaCl solution and connected with calomel electrodes by means of salt bridges; and glass-insulated carbon electrodes (tip diameter, 50 μm). The carbon electrodes had a DC resistance of 10-50 MOhm and were really nonpolarizable when the input resistance of the DC amplifier exceeded 1 GOhm, which corresponds to the data reported by Roth et al. (1966) and by Starrenberg and Burger (1982). Electrochemical sensitivity of the carbon electrodes did not result in distortion of the negative potential accompanying the SD wave. Evidence was obtained in a special series of experiments when SD waves were recorded by calomel and carbon electrodes simultaneously (unpublished data). In experiments on immobilized animals, a calomel electrode, placed on the soft tissues of the head, was used as a reference electrode. In experiments on freely moving animals a silver ball, covered with silver chloride and placed on the back bones of the skull, served as a reference electrode.

Direct current potential and EEG changes were recorded by a six-channel high-impedance input amplifier and recorded with a polygraph. The overall bandpass of the recording system ranged from 0 to 100 Hz.

Electrical Stimulation

Electrical stimulus trains were delivered in a monopolar manner through glass-insulated carbon electrodes, since they successfully combine the properties of weakly polarizable electrodes for the brain electrical activity recording and metallic electrodes for ES. The resistance of the carbon electrodes to the alternating current when the pulse duration was 0.3 ms did not exceed 10 KOhm. In experiments with direct thalamic or hippocampal ES double-barreled carbon electrodes were used. One barrel served as a stimulating electrode and the other as a recording one.

The stimulus trains consisted of monophasic squared pulses with a duration of 200-300 μs and an amplitude of 50-1000 μA. The frequency of stimulation varied from 10 to 500 Hz.

Apparatus

The animal's behavior was tested in a box of the "open field" type with dimensions 750×750 mm. The floor of the box consisted of 25 squares, which had built-in capacity detectors to register the location of the rat in the box. The information from the detectors was processed by computer, which enabled: (1) the animal moving traces to be plotted; (2) the locomotor activity to be evaluated from the number of squares traversed per unit time; and (3) the time the animal remained in any place of the box to be calculated. The box was electrically shielded, which enabled the electrical activity of the brain to be recorded simultaneously with the animal's movement.

Results

Different types of rat brain electrophysiological responses to short cortical ES with different frequencies are shown in Fig. 1. Parietal cortical ES with a frequency of 50 Hz and a threshold intensity for the cortical SD also provoked a short seizure afterdischarge (AD) in the brain. With an ES over threshold the afterdischarge became longer and was often accompanied by hippocampal SD and rather seldom by thalamic SD. This sequence of events corresponds to data showing a good probability of SD occurrence in deep brain structures on the basis of an intensive AD (Siegfried et al. 1977; Bureš and Burešová 1981; Koroleva and Bureš 1982). In our experiments cortical ES with a frequency of 50 Hz resulted in an SD wave in the hippocampus in 50% of cases and in the thalamus in 15% of cases. The dynamics of local and projected responses changed substantially with the increase in cortical ES frequency up to 200-500 Hz (Fig. 1b). In this case a very regular occurrence of the SD wave in the thalamus was observed under a definite ES intensity. Surprisingly, the thalamic SD threshold was considerably lower than the AD threshold.

Electroencephalogram control, carried out simultaneously with DC slow potential shift recording, confirmed a total absence of seizure activity in the cortex and subcortical structures, when the intensity and duration of the high-frequency cortical ES was effective in causing thalamus SD (Fig. 2). Nembutal anesthesia increased the thresholds, but did not prevent SD in the thalamus. Maximum-amplitude SD waves developed as a result of the shortest ES, 0.1 s long (Fig. 3). With an increase of ES duration up to 1-2 s and longer, the amplitude of the thalamus SD waves decreased and stabilized at the level of 5-6 mV. Thus an increase in ES duration gave rise to the activation of inhibiting processes, preventing the full development of SD. A train of 10-20 pulses was optimal for the induction of thalamic SD, though sometimes 3-5 pulses were enough. When the current intensity exceeded the cortical SD thresholds by a factor of 2-3 and equaled 300-500 μA, the probability of

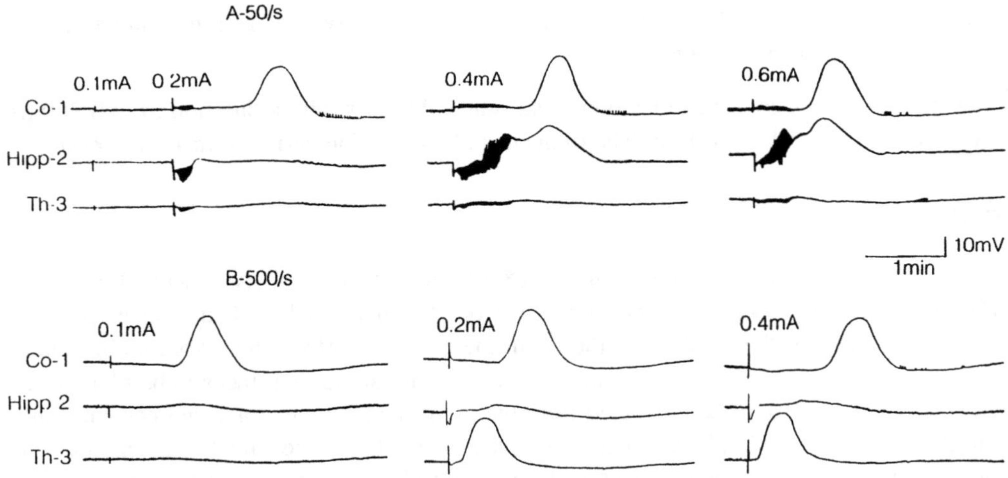

Fig. 1 A,B. Electrophysiological responses of rat brain to short high-frequency electrical stimulation of the parietal cortex. The responses to 50-Hz stimulus trains (A) and 500-Hz stimulus trains (B) are from the same animal. DC slow potentials were recorded monopolarly in cortex, hippocampus, and thalamus. Repetitive cortical ES was produced at 15-min intervals. The stimulus train consisted of 20 squared pulses with durations of 0.3 ms. The intensity of brain stimulation was incremented in 100- to 200-μA steps

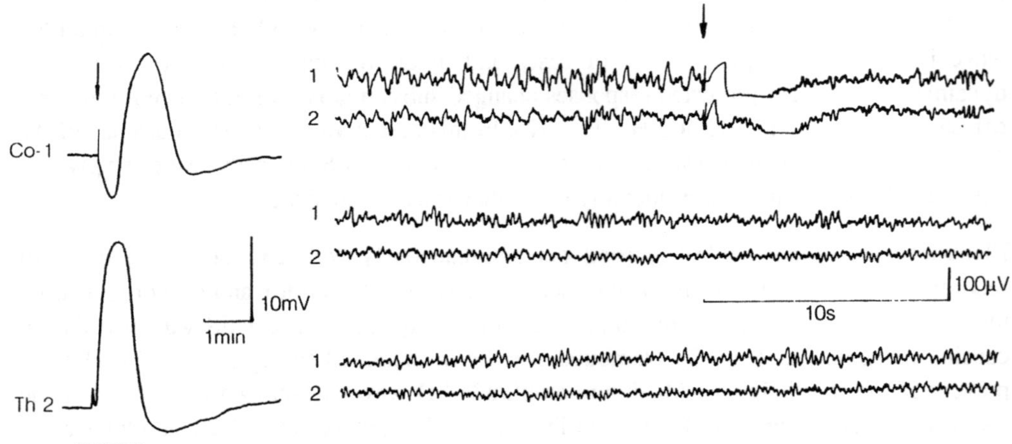

Fig. 2. Electroencephalogram control of seizure activity absence after high-frequency cortical ES, effective for thalamic SD incidence. DC slow potential (*left*) and EEG (*right*) were recorded monopolarly from the same cortical and thalamic electrodes. EEG recordings were made continuously during the period indicated by the *bar* under the thalamic slow potential recording. ES is indicated by *arrows*. The electrical stimulus consisted of 10 pulses with a duration of 0.3 ms. The frequency of brain stimulation was 200 Hz and the intensity was 300 μA

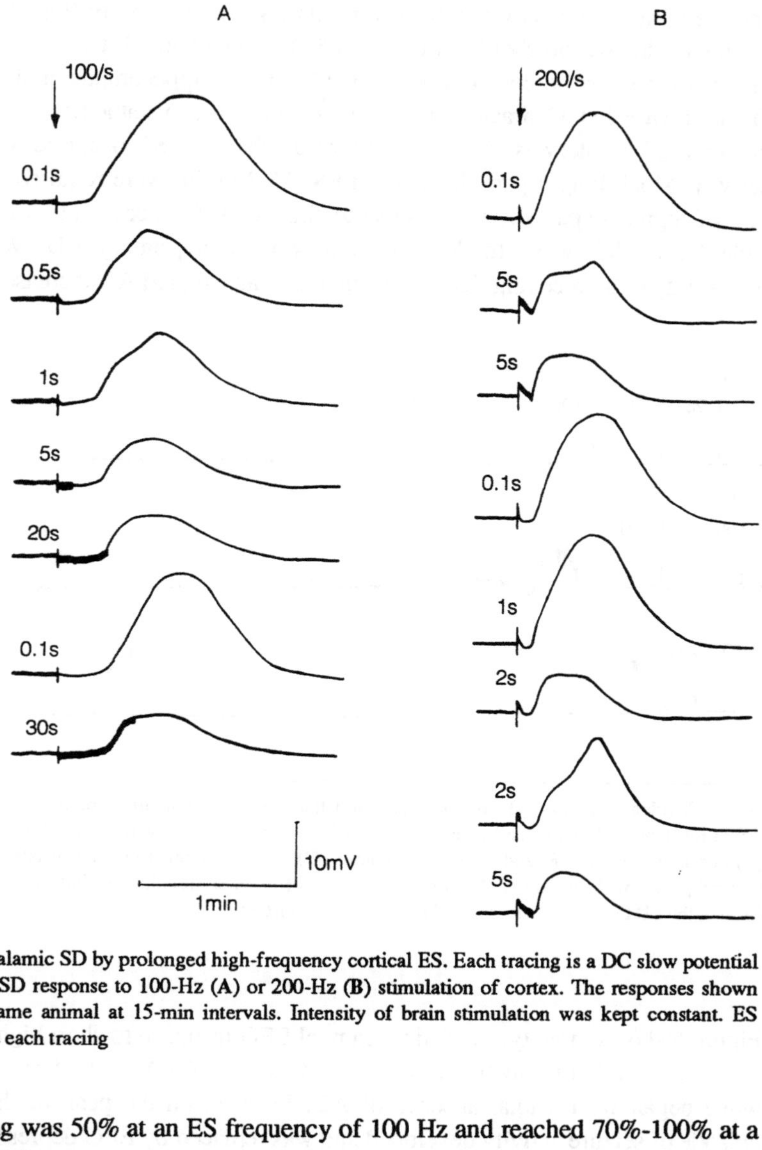

Fig. 3 A,B. Blocking of thalamic SD by prolonged high-frequency cortical ES. Each tracing is a DC slow potential recording of the thalamic SD response to 100-Hz (A) or 200-Hz (B) stimulation of cortex. The responses shown were obtained from the same animal at 15-min intervals. Intensity of brain stimulation was kept constant. ES duration is indicated above each tracing

thalamic SD occurring was 50% at an ES frequency of 100 Hz and reached 70%-100% at a frequency of 200-500 Hz.

In contrast to in the thalamus, thresholds of the synaptically induced hippocampal SD were three to five times higher than those of cortical SD. Over the whole range of frequencies the probability of SD occurrence in the hippocampus was no greater than 40%-50% and there was a high correlation of hippocampal SD and seizure AD.

To answer the question of how closely the listed differences in SD dynamics in the deep structures are connected with their own distinctive features, experiments were carried out with direct high-frequency ES of the hippocampus and the thalamus. The deep structures and the cortex were shown to have very high SD susceptibility (Fig. 4). In the different experiments the SD thresholds varied from 50 to 300 μA in all structures, giving an average of 184 ± 25 μA for the cortex, 200 ± 38 μA for the hippocampus, and 153 ± 29 μA for the thalamus (n = 13). Characteristic for the thalamus was a stable SD threshold under repeated thalamic ES at intervals of 5-6 min. All of the thalamic SD occurred with no sign of seizure activity. Much longer periods between ES (15-20 min) were required for regular reproduction of hippocampal SD. In response to short high-frequency hippocampal ES it was possible for an SD wave to develop there without a primary AD. With repeated ES the probability of a preceding SD afterdischarge or an isolated AD increased.

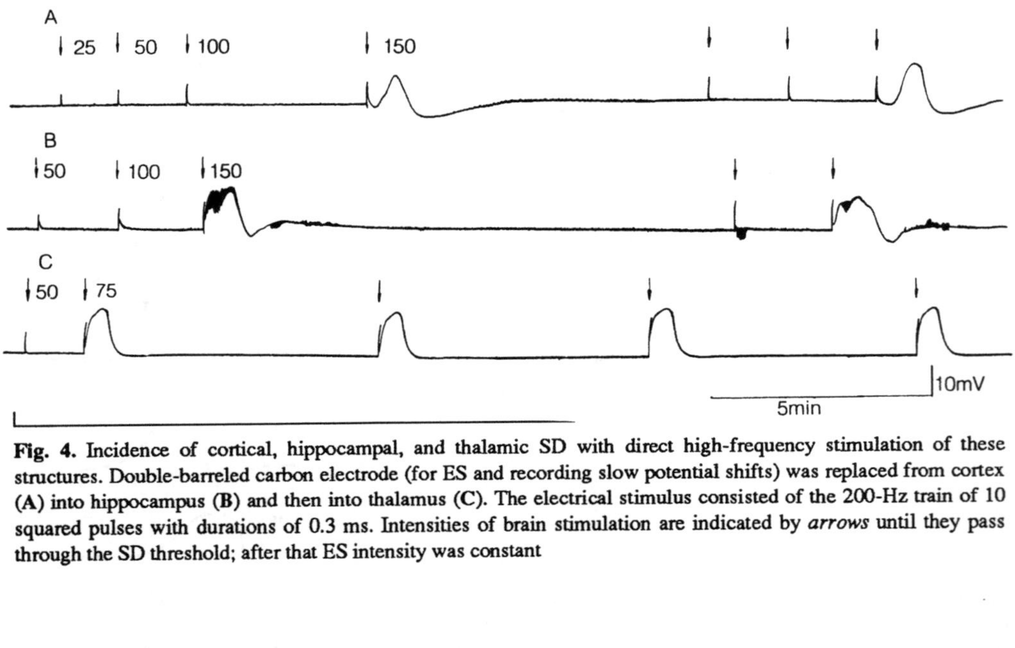

Fig. 4. Incidence of cortical, hippocampal, and thalamic SD with direct high-frequency stimulation of these structures. Double-barreled carbon electrode (for ES and recording slow potential shifts) was replaced from cortex (A) into hippocampus (B) and then into thalamus (C). The electrical stimulus consisted of the 200-Hz train of 10 squared pulses with durations of 0.3 ms. Intensities of brain stimulation are indicated by *arrows* until they pass through the SD threshold; after that ES intensity was constant

Figure 5 shows three types of hippocampal EEG response to short high-frequency ES effective (A, B) and not effective (C) for the induction of SD. In the first case (Fig. 5A) the SD wave began with a total absence of AD. In 20 s (on the peak of the SD wave) a weak episode of seizure activity developed and was replaced by EEG depression. A secondary AD appeared clearly in 160 s on the background of a deep EEG depression. This characteristic burst of paroxysmal activity 10-12 s long always accompanied the hippocampal SD and was recorded in many structures. In the second case (Fig. 5B), a primary AD 18 s long developed in the ascending phase of the SD wave and was replaced by EEG depression in the descending phase. The secondary burst of paroxysmal activity also occurred after a prolonged silent period. The fact that this type of seizure activity is a consequence of the hippocampal SD rather than a primary AD is illustrated in Fig. 5C. If an isolated AD (without an SD wave) occurred, it was never accompanied by secondary paroxysmal activity.

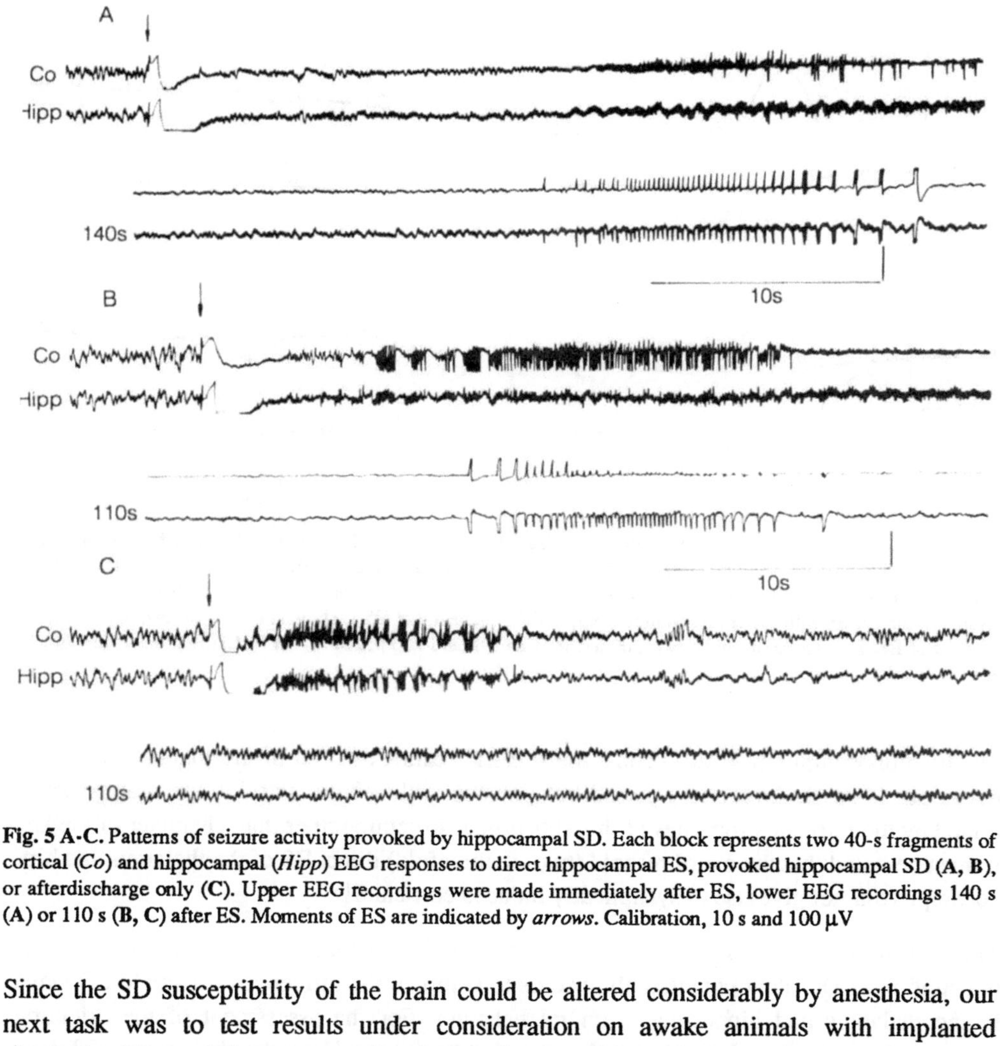

Fig. 5 A-C. Patterns of seizure activity provoked by hippocampal SD. Each block represents two 40-s fragments of cortical (*Co*) and hippocampal (*Hipp*) EEG responses to direct hippocampal ES, provoked hippocampal SD (**A, B**), or afterdischarge only (**C**). Upper EEG recordings were made immediately after ES, lower EEG recordings 140 s (**A**) or 110 s (**B, C**) after ES. Moments of ES are indicated by *arrows*. Calibration, 10 s and 100 μV

Since the SD susceptibility of the brain could be altered considerably by anesthesia, our next task was to test results under consideration on awake animals with implanted electrodes. Figure 6 shows parts of such experiments with cortical, thalamic, and hippocampal ES. It is evident that under given circumstances we obtained the same correlations as we observed in experiments on anesthetized animals. Depending on the parameters of cortical stimulation, SD occurred only in the cortex or SD was also Synaptically induced in the thalamus with no signs of seizure activity (Fig. 6A, B). With direct high-frequency thalamic ES the SD occurred readily in the absence of any AD (Fig. 6C). But the same hippocampal ES triggered SD without preceding AD only in 31% of cases (total number of ES = 206). In other cases the SD was preceded by short (10-30 s) AD (47%) or an isolated AD was observed. Within 2-2.5 min after the hippocampal SD a characteristic secondary burst of paroxysmal discharges developed, which was better expressed in other structures than in the dorsal hippocampus (Fig. 6D). The SD thresholds with the direct thalamic and hippocampal stimulation of awake animals did not differ significantly from the thresholds obtained in anesthetized animals and were on average 150 ± 25 μA for the thalamus and 330 ± 32 μA

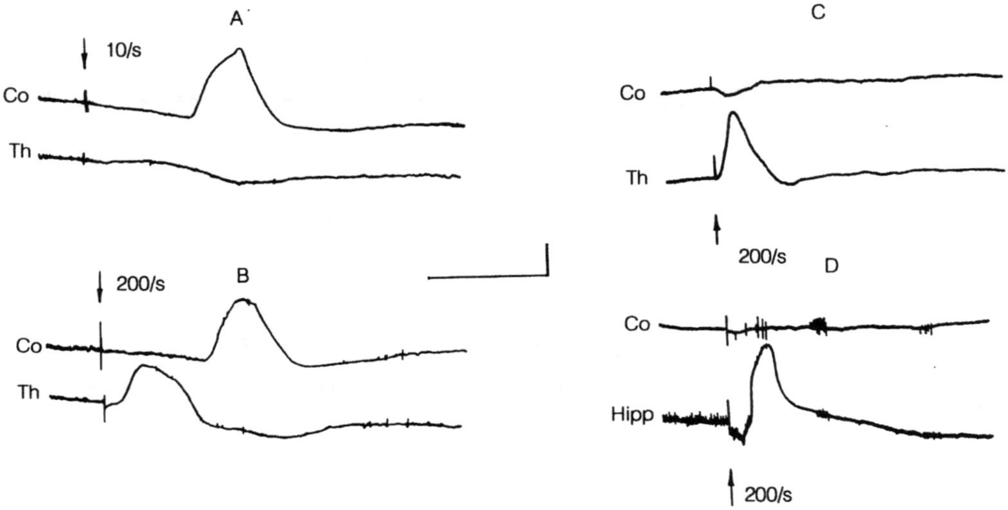

Fig. 6. Typical examples of SD initiation after subseizure ES of cortex (**A, B**), thalamus (**C**), and hippocampus (**D**) in freely moving rats. DC slow potential recordings were made by carbon electrodes in cortex (*Co*), thalamus (*Th*), and hippocampus (*Hipp*). Besides the stimulated area DC potentials were controlled in the thalamus with cortical ES and in the cortex with thalamic and hippocampal ES. Frequency of ES is shown near the *arrows* indicating the moment of ES. Amplitude of pulses was 500 μA in **A** and **B**, and 200 μA in **C** and **D**. Calibration, 1 min and 10 mV

for the hippocampus (n = 20). Thus, the specific features of SD in the thalamus and hippocampus (threshold, recovery period, relationship with seizure activity) are mainly determined by the intrinsic properties of this subcortical structure rather than character of the ES. It must be noted that the overthreshold (two- to threefold) direct ES of the thalamus and hippocampus did not give rise to synaptically induced cortical SD.

The possibility of isolated provocation of SD in deep structures with the simultaneous registration of DC slow potential allowed us to compare the influence of a single SD wave in the thalamus and hippocampus on rat behavior. For the assessment of the orienting response the rats were placed in the "open field" for 30 min/day over the course of 3 days before the ES. Beginning on the 4th day, ES of the hippocampus or thalamus, which caused the SD, was carried out. The appearance of SD in the hippocampus resulted in the animal stopping its current activity and starting to move rapidly in the box. The movement was accompanied by frequent standing on hind legs. In the 1st and especially in the 3rd min after the ES, bouts of shaking, a well-known behavioral component of hippocampal SD (Siegfried et al. 1977), developed. Figure 7 shows the traces of rat walking during 5 min of the orienting response (Fig. 7A) and during 5 min after the hippocampal ES, which triggered an SD wave in an awake (Fig. 7B) and sleeping (Fig. 7C) animal. As can be seen, the locomotor activity which accompanied the hippocampal SD was independent of the background activity level and had a pronounced pre-wall character. The increase in locomotor activity began in the 1st min after the ES and reached its maximum within 2-3 min. Locomotor excitation disappeared within 6-7 min and was often replaced by quietness.

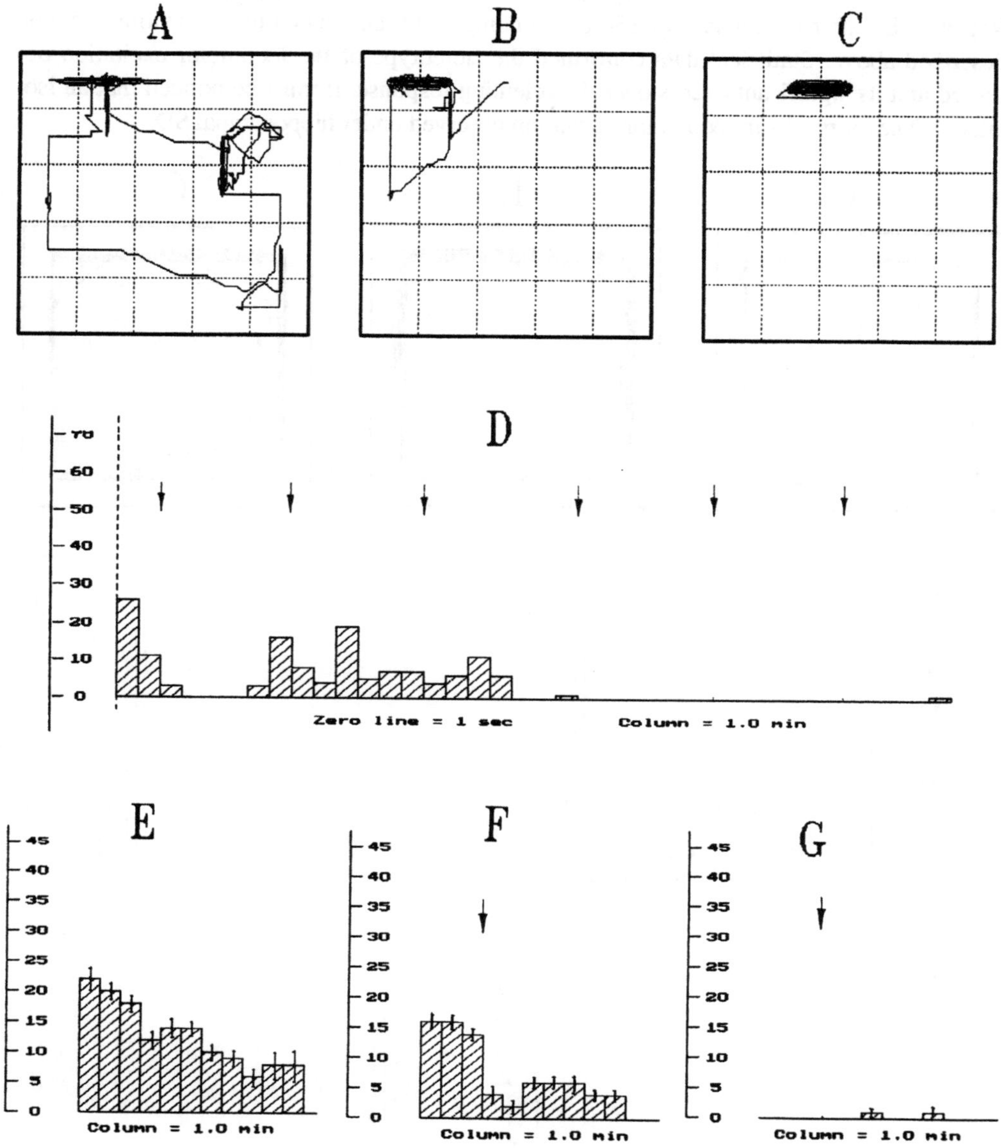

Fig. 7 A-F. Locomotor hyperactivity of rats provoked by hippocampal SD. A-C Traces of rat moving in an open field for 5 min during orinenting response (**A**) and after ES evoked hippocampal SD (**B, C**). **D** Dynamics of locomotor activity during one experiment with twice repeated hippocampal ES evoking SD in this structure on an active and tranquil background; moments of ES are indicated by *arrows*. *Columns* indicate the number of squares crossed by the rat in 1 min. **E** Average locomotor activity during 8 min of orienting response (for first days), n = 30. **F** Average locomotor activity for 3 min before and 7 min after ES evoked hippocampal SD, n = 80. Ordinate is the same as in **D**

Repeated ES after the interval of 15-20 min completely reproduced the behavioral reaction described above. Statistical data confirmed the stereotype of the locomotor excitation observed and its significant excess over the orienting response. It must be noticed that an isolated AD never results in locomotor activation observed under hippocampal SD.

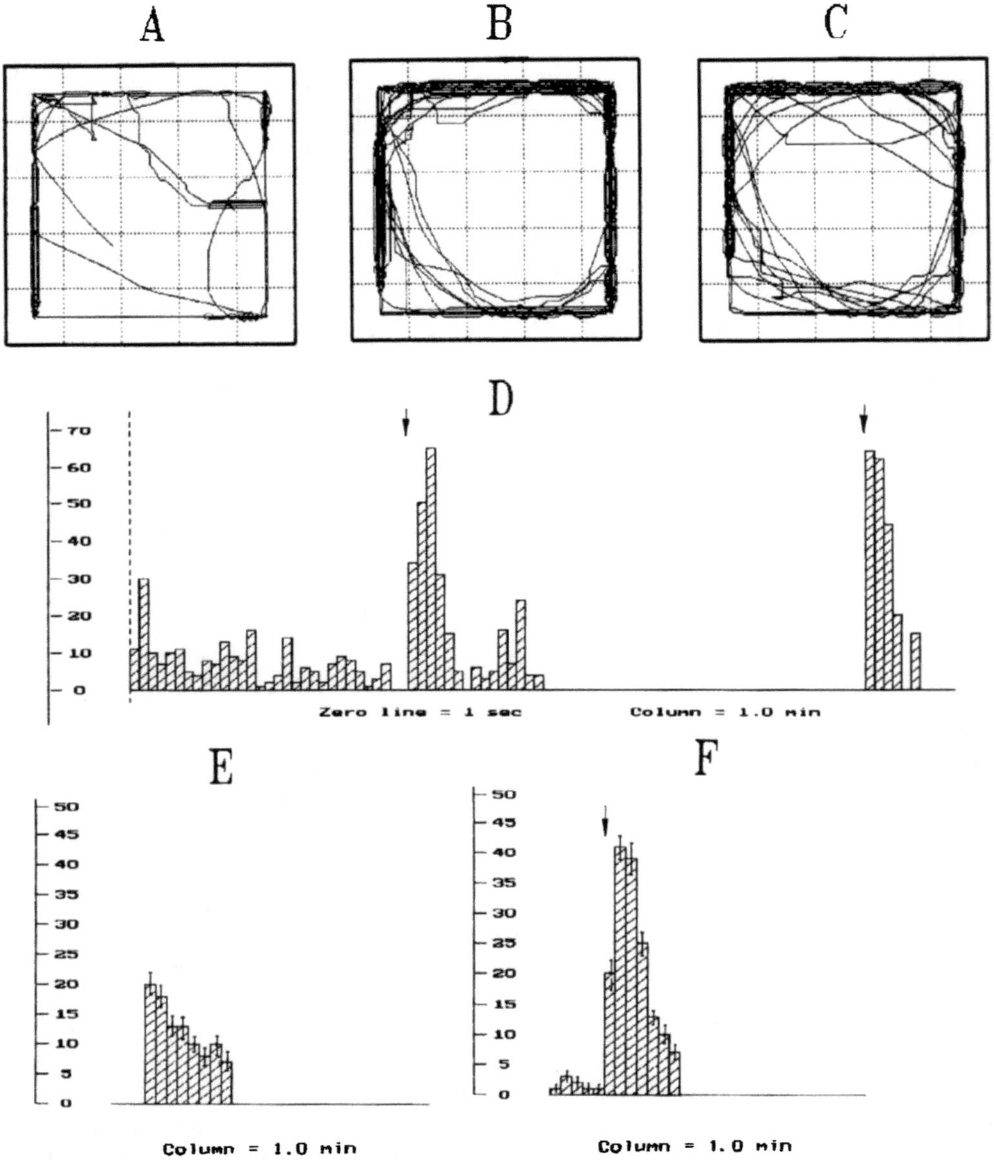

Fig. 8 A-G. Behavioral effect of thalamic SD. **A-C** Three examples of rat moving in an open field for 5 min after thalamic ES induced SD. **D** Dynamics of locomotor activity during one experiment with regular thalamic SD; moments of ES are indicated by *arrows*. *Columns* indicate the number of squares crossed by the rat in 1 min. **E** Average locomotor activity during 10 min of orienting response, $n = 30$. **F,G** Average locomotor activity before and after thalamic SD evoked by ES (*arrows*) on the active backgroung (**F**) and on the relaxed state (**G**)

Figure 8 shows the corresponding characteristics of the locomotor activity, induced by the thalamic SD. Firstly it is necessary to note the absence of any specific locomotor excitation in connection with stimulation and SD development in the thalamus. The character of the reaction was determined by the level of background locomotor activity. When the orienting response was not damped, the thalamic SD effectively reduced the mobility of the animal (Fig. 8D, F), but dit not completely stop the current activity (grooming, eating). The SD, elicited in the thalamus on a nonactive background, never resulted in the activation of the animal (Fig. 9D, G).

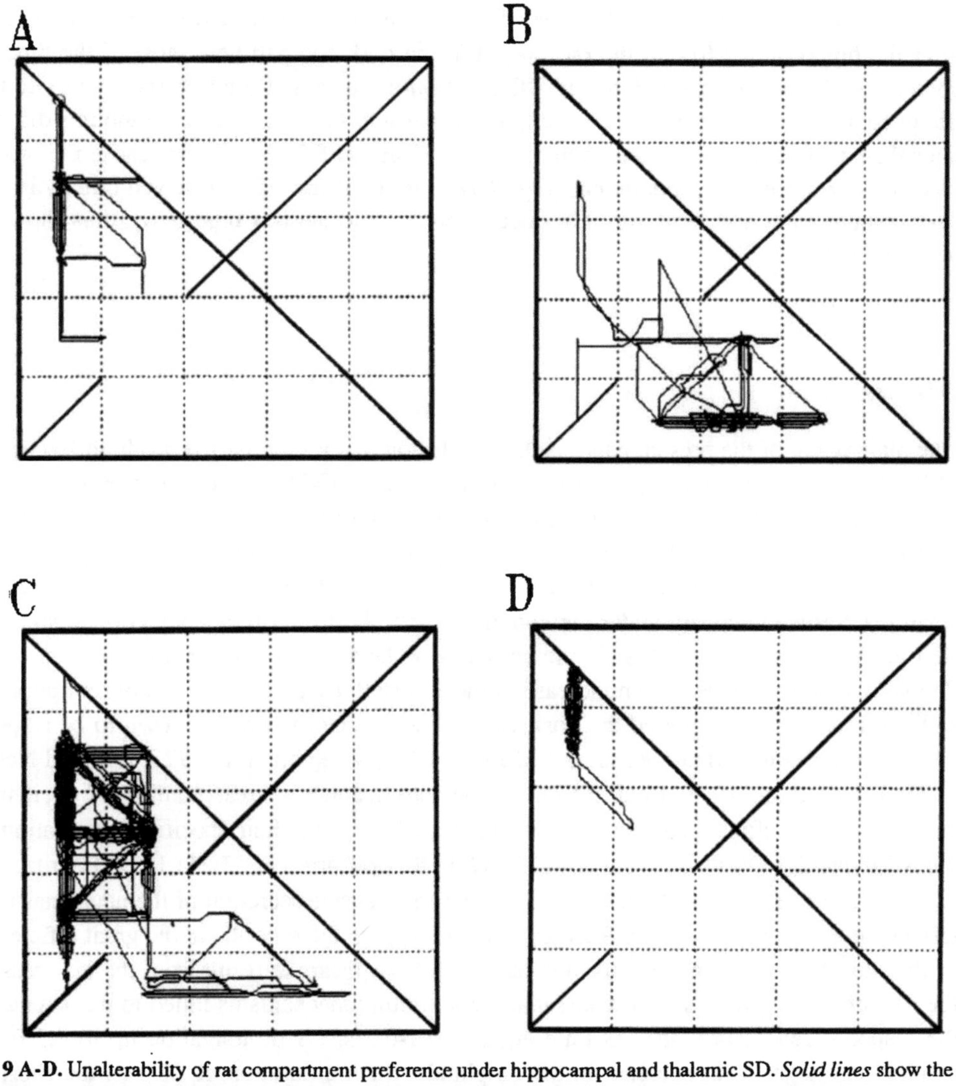

Fig. 9 A-D. Unalterability of rat compartment preference under hippocampal and thalamic SD. *Solid lines* show the modification of an open field to the Skinner box. Rats were allowed to move freely between left (*dark*) and low (*light*) departments. Other departments were inaccessible. Traces of rat moving during 10 min after habituation (A), after the introduction in the left (dark) compartment of sound (B), hippocampal SD (C), and thalamic SD (D)

In our experiments we never noticed that the animals avoided entering a region of an open field previously associated with SD initiation. This fact led us to the assumption that SD in the structures investigated is not aversive. To test this proposition two compartments with a different degree of illumination were selected in the box. Rats were able to move between them freely. After damping the orienting response (over the course of 3 days), all the rats preferred to stay in the dark compartment (Fig. 9A). Then the animals were divided into three groups in order to produce in the preferred compartment an additional influence such as sound and hippocampal and thalamic SD. Each rat was tested in ten experiments with two to three repeated SD elicitations. After two or three sound presentations rats changed their preference, staying in the illuminated compartment all the time (Fig. 9B). During and after the hippocampal SD all the rats stayed in the dark compartment most of the time in spite of the locomotor excitation (Fig. 9C). Only infrequent and brief journeys out into the illuminated compartment were observed. When put into the box repeatedly the rats did not alter their preference for the dark compartment. Thalamic SD also did not cause the rats to change their preferred compartment (Fig. 9D). With thalamic ES rats stayed quietly in the dark compartment and went into it at once when put into the box repeatedly, thus showing no aversiveness to SD.

Discussion

The main result of the present paper is the conclusion that a short (0.1-0.2 s) high-frequency (100-500 Hz) electrical stimulation with an intensity of 50-300 μA easily provokes an SD wave in the thalamus and hippocampus. The SD can develop in these structures not only as a result of direct ES but also in response to a synaptic signal from the cortex. Synaptically triggered SD occurred most regularly in the thalamus without being accompanied by any seizure activity. Obviously under the conditions of a short high-frequency cortical stimulation in the thalamus as well as in the cortex (Koroleva 1989), an effective summation of depolarization processes and an increase in the extracellular potassium ion concentration up to the critical values triggered the positive feedback system which is known to be responsible for the SD initiation (Matsuura and Bureš 1971; Vyskočil et al. 1972; Lux and Neher 1973; Prince et al. 1973; Reid 1987). Taking into account a great number of excitatory aminoacid corticofugal projections (Cotman et al. 1987) and their specific participation in the SD dynamics (Van Harreveld and Fifková 1970; do Carmo and Leão 1972; Lauritzen et al. 1988; Marranes et al. 1988), one can assume an intensive secretion of the neurotransmitters in the areas of the synaptic SD focus formations. On the other hand the great efficiency of the short high-frequency ES for provoking the SD appears to result from the inertness of the SD-inhibiting processes. It seems that the inhibiting mechanisms failed to participate in the response reactions if the ES time equaled 0.1-0.2 s. ES prolongation up to 0.5-0.2 s resulted in a significant decrease in SD amplitude. Such a duration of ES was too short to activate successfully the nonsynaptic pumping mechanisms of the SD blockade (Koroleva and Bureš 1980; Koroleva and Gorelova 1983). That is why it may be supposed that inhibitory synaptic processes mainly participate in the decrease of SD amplitude. Nonsynaptic mechanisms are known to be responsible for the establishment of sustained steady potential (ceiling level) due to very prolonged ES (Futamachi et al. 1974; Heinemann and Lux 1977; Somjen 1979). In these conditions the development of SD may be only a result of failure of ionic homeostasis in the CNS (Nicholson 1980). In general, our data about a high susceptibility of nervous tissue to SD not only in the stimulated area but also in the synaptically connected areas show that SD cannot be regarded only as a pathological phenomenon.

Since SD readily occurs in the thalamus under the cortical ES, it is interesting to go back to the discussion of the 1940s and 1950s about the identity of SD and spreading suppression (Dusser de Barenne and McCulloch 1941; Sloan and Jasper 1950). The main difference was the absence of the seizure activity during the spreading suppression and the involvement of the thalamocaudate system in this reaction. Evidently the motor response suppression induced by cortical ES could be a result of thalamic rather than cortical SD. In our experiments with freely moving rats, isolated thalamic SD had a clear suppressive effect on their behavior. The state of rats may be easily interpreted on the basis of the classic concept of SD as a functional ablation phenomenon (Bureš et al. 1974). Extensive electrophysiological research by Bureš and his coworkers on remote effects of K-induced thalamic SD have shown a great reduction of unit and EEG activity not only in the thalamus but in many other structures, decrease of cortical excitability, and other consequences of cortical deafferentation. To a considerable degree the lack of research about the behavioral consequences of thalamic SD is due to the side effects of KCl application in subcortical centers because of its backward spread or diffusion into the neighboring structures and possible initiation of local seizures. Our method of triggering thalamic SD by short high-frequency ES gave us the opportunity to avoid these difficulties. Under regular thalamic SD, initiated at 6- to 15-min intervals, there were no such complications and rats remained tranquil throughout the experiment.

In contrast to the thalamus, hippocampal SD induced by high-frequency ES was closely connected with the seizure AD which developed simultaneously or preceded SD. Isolated AD (without SD) was elicited only occasionally. A secondary AD was shown to develop only when SD appeared in the hippocampus, which is in agreement with the data reported earlier about different signs of hyperactivity following the SD wave (Bureš et al. 1974; Huston and Holzhäuer 1988). In many investigations with high-frequency ES of the hippocampus and connected structures, the pattern and dynamics of AD were studied without direct current recording. The secondary (or recurrent, or rebound) discharge observed in these studies is usually interpreted as a manifestation of long-loop reverberating activity in neural circuits (Leung 1987). A strict dependence between hippocampal SD and the similar secondary AD demonstrated in our report points to the special role of SD waves in recurrent paroxysmal episodes.

While interpreting the behavioral effects of hippocampal SD, one usually takes into account the inhibitory nature of connections of the hippocampus with associated brain centers and projection of the hippocampal paroxysm to connected areas (Bureš et al. 1974, 1984; Huston et al. 1974). In our experiments short primary AD was shown to influence weakly the behavior when it appeared as an isolated event. Therefore, locomotor hyperactivity observed evidently resulted from hippocampal SD itself; however, secondary discharge might add something to the excitatory effect of SD. A low threshold of SD triggering in the hippocampus can explain a characteristic similarity of behavioral effects after weak ES and ablation of the hippocampus (Vinogradova 1975).

To interpret the behavioral effects of SD, its motivational properties must be taken into account. In some papers attempts were made to resolve whether SD is aversive or not. It was shown that K-induced SD in vestibular nuclei produced taste aversion (Burešová et al. 1982). The data obtained for cortical SD was contradictory because of its possible influence on learning (Freedman and Whitman 1972; Winn et al. 1977; Davis et al. 1983; Bureš et al. 1984). In our experiments the rats did not change the compartment preference after 20-45 SD waves in the thalamus and hippocampus in spite of a pronounced difference in the locomotor SD effects. As to the hippocampal SD, the data obtained do not allow a definitive conclusion to be drawn about its nonaversiveness, because the hippocampal SD influences learning (Bureš et al. 1974; Kapp and Schneider 1971; Schneider and Simson 1983). Since thalamic SD is not known to influence the learning of passive avoidance (Burešová et al. 1965), our data testify to the fact that thalamic SD is really not aversive.

References

Aquino-Cias J, Estevez O, Aneiros-Riba R (1971) Efectos de la depresion propagada cortical sobre las proyecciones de focos penicilinicos en la corteza cerebral de la rata. In: Resumenes del III Seminario Cientifico del CNIC. La Habana, pp 197-198

Bureš J, Burešová O (1981) Cerebral $[K+]_e$ increase as an index of the differential susceptibility of brain structures to terminal anoxia and electroconvulsive shock. J Neurobiol 12: 211-220

Bureš J, Burešová O, Weiss T (1960) Functional consequences of hippocampal spreading depression. Physiol Bohemoslov 9: 219-227

Bureš J, Burešová O, Křivánek J (1974) The mechanism and applications of Leão's spreading depression of electroencephalographic activity. Academic, New York

Bureš J, von Schwarzenfeld I, Brožek G. (1975) Blockage of cortical spreading depression by picritoxin foci of paroxysmal activity. Epilepsia 16: 111-118

Bureš J, Burešová O, Křivánek J (1984) The meaning and significance of Leão's spreading depression. An Acad Brasil Ciênc 56: 385-400

Burešová O, Fifková E, Ross RB (1965) The effect of thalamic spreading depression on acquisition and consolidation of the passive avoidance reaction in rats. Act Nerv Super 7: 162-163

Burešová O, Semenov LV, Bureš J (1982) Electrical stimulation or chemical blockade of vestibular nuclei can serve as the unconditioned stimulus in the conditioned taste aversion paradigm. Acta Biol Acad Sci Hung 33: 139-148

Cotman CW, Monagham DT, Ottersen OP, Storm-Mathisen J (1987) Anatomical organization of excitatory amino acid receptors and their pathways. TINS 10: 273-280

Davis JL, Burešová O, Bureš J (1983) Cortical spreading depression and conditioned test aversion: an attempt to resolve a controversy. Behav Neural Biol 37: 338-343

de Luca AM, Pivik RT, Chorover SL (1977) Electroconvulsive shock: effects on sleep and cortical steady potential in the rat. Physiol Behav 18: 997-1003

do Carmo RJ, Leão AAP (1972) On the relation of glutamic acid and some allied compounds to cortical spreading depression. Brain Res 39: 515-518

do Carmo RJ, Martins-Ferreira H (1984) Spreading depression of Leão probed with ion-selective microelectrodes in isolated chick retina. An Acad Brasil Ciênc 56: 401-421

Dusser de Barrene JG, McCulloch WS (1941) Suppression of motor response obtained from area 4 by stimulation of area 4s. J Neurophysiol 4: 311-323

Fifková E, Maršala J (1967) Stereotaxic atlases for the cat, rabbit and rat. Electrophysiological methods in biological research. Czechoslovaiaan Academy of Sciences, Prague, pp 653-731

Freedman NL, Whitman RD (1972) Aversiveness of functional decortication and performance deficit. J Life Sci 2: 73-79

Futamachi KJ, Mutani R, Prince DA (1974) Potassium activity in rabbit cortex. Brain Res 75: 5-25

Guedes RCA (1984) On some conditions that influence cortical spreading depression. An Acad Brasil Ciênc 56: 445-455

Guedes RCA, de Azeredo FAM, Hicks TP, Clark RJ, Tashiro T (1987) Opioid mechanisms involved in the slow potential change and neuronal refractoriness during spreading depression. Exp Brain Res 69: 113-118

Heinemann U, Lux HD (1977) Ceiling of stimulus-induced rises of extracellular potassium concentration in the cerebral cortex of cat. Brain Res 120: 231-249

Heinemann U, Lux HD, Gutnick MI (1977) Extracellular free calcium and potassium during paroxysmal activity in the cerebral cortex of the cat. Exp Brain Res 27: 237-243

Huston JP, Holzhäuer M-S (1988) Behavioral and electrophisiological effects of intracranial applied neuropeptides with special attention to DC slow potential changes. Ann N Y Acad Sci 525: 375-390

Huston JP, Siegfried B, Ornstein K, Waser PG, Borbely AA (1974) Eating elicited by spreading depression or electrical stimulation of the hippocampus and neocortex: a common cause. Brain Res 78: 164-168

Kapp BS, Schneider AM (1971) Selective recovery from retrograde amnesia produced by hippocampal spreading depression. Science 173: 1149-1151

Koroleva VI (1989) Cortical spreading depression synaptically induced by short high-frequency electrical stimulation of the rat brain (in Russian). Neurofiziologia 21: 789-796

Koroleva VI, Bureš J (1980) Blockade of cortical spreading depression in electrically and chemically stimulated areas of cerebral cortex in rats. Elektroencephalogr Clin Neurophysiol 48: 1-15

Koroleva VI, Bureš J (1982) Stimulation induced recurrent epileptiform discharges block cortical and subcortical spreading depression in rats. Physiol Bohemoslov 31: 385-400

Koroleva VI, Gorelova NA (1983) Properties of cortical spreading depression in different phases of cyclic excitation of rat neocortex (in Russian). Neurofiziologia 15: 226-234

Koroleva VI, Vinogradova LV (1990) Spreading depression in the thalamus, hippocampus and caudate nucleus triggered by electrical stimulation of parietal cortex of the rat (in Russian). Neurofiziologia 22: 36-44

Koroleva VI, Oitzl M-S, Bureš J (1985) Threshold of synaptically elicited cortical spreading depression: drug induced changes in rats. Electroencephalogr Clin Neurophysiol 60: 55-64

Lauritzen M, Rice ME, Okada Y, Nicholson C (1988) Quisqualate, kainate and NMDA can initiate spreading depression in the turtle cerebellum. Brain Res 475: 317-327

Leão AAP (1944) Spreading depression of activity in the cerebral cortex. J Neurophysiol 7: 359-390

Leão AAP (1951) Slow voltage variation of cortical spreading depression of activity. Electroencephalogr Clin Neurophysiol 3: 315-321

Leão AAP (1963) On the spread of spreading depression. In: Brazier MA (ed) Brain function, vol 1. University of California Press, Los Angeles, pp 73-85

Leung L-WS (1987) Hippocampal electrical activity following local tetanization. 1. Afterdischarges. Brain Res 419: 173-187

Lux HD, Neher E (1973) The equilibration time course of $[K+]_e$ in cat cortex. Exp Brain Res 17: 190-205

Machek J, Ujec E, Pavlík V, Horák F (1977) Interhemispheric field potentials, spreading depression and kindling. Neurosci Lett 4: 337-341

Marranes R, Willems R, De Prins E, Wauquir A (1988) Evidence for a role of the *N*-methyl-*D*-aspartate (NMDA) receptor in cortical spreading depression in the rat. Brain Res 457: 226-240

Matsuura T, Bureš J (1971) The minimum volume of the depolarized neural tissue required for the triggering cortical spreading depression in rat. Exp Brain Res 12: 238-249

Nicholson C (1980) Dynamics of the brain cell microenvironment. Neurosci Res Prog Bull 18: 177-322

Oitzl M-S, Huston JP (1984) Electroencephalographic spreading depression and concomitant behavioral changes induced by intrahippocampal injection of ACTH and d-Ala^2-metenkephalinamide in the rat. Brain Res 308: 33-42

Prince DA, Lux HD, Neher E (1973) Measurement of extracellular potassium activity in cat cortex. Brain Res 50: 489-495

Purpura DP (1963) Review and Critique. In: Brazier MA (ed) Brain function. University of California Press, Los Angeles, pp 281-320

Reid KH (1987) Potassium translocation and spreading depression induced by electrical stimulation of the brain. Exp Neurol 97: 345-364

Roth J, Cullen HM, Milstein V (1966) The use of carbon electrodes for chronic cortical recording. Electroencephalogr Clin Neurophysiol 21: 611-615

Schneider AM, Simson PE (1983) Spreading depression: a behavioral analysis. In: Deutch JA (ed) The physiological basis of memory Academic, New York, pp 121-138

Siegfried B, Huston JP (1976) Properties of spreading depression-induced consummatory behavior in rats. Physiol Behav 18: 841-851

Siegfried B, Shibata M, Huston JP (1977) Electrophysiological concomitants of eating induced from neocortex and hippocampus by electrical stimulation and injection of KCl or norepinephrine. Brain Res 121: 97-112

Sloan N, Jasper H (1950) Identity of spreading depression and "supression". Electroencephalogr Clin Neurophysiol 2: 59-78

Somjen GG (1979) Extracellular potassium in the mammalian central nervous System. Annu Rev Physiol 41: 159-177

Starrenberg AJM, Burger GC (1982) Carbon fiber as an electrode material. IEEE Trans Biomed Eng [BME-29] 5: 352-355

Stringer JL, Lothman EW (1989) Maximal dentate gyrus activation: characteristics and alterations after repeated seizures. J Neurophysiol 62: 136-143

Van Harreveld A, Fifková E (1970) Glutamate release from the retina during spreading depression. J Neurobiol 2: 13-29

Van Harreveld A, Stamm JS (1953) Spreading cortical convulsions and depression. J Neurophysiol 26: 352-366

Vinogradova OS (1975) Hippocampus and memory. Nauka, Moscow

Vyskočil F, Křiž N, Bureš J (1972) Potassium-selective microelectrodes used for measuring the extracellular brain potassium during spreading depression and anoxic depolarization in rat. Brain Res 39: 255-259

Whieldon JA, Van Harreveld A (1950) Cumulative effects of minimal cortical stimulation. Electroencephalogr Clin Neurophysiol 2: 49-57

Winn FJ, Todd GE, Elias JW (1977) Cortical spreading depression-induced aversion to saccharin at two levels of KCl: electroencephalographic verification. Behav Biol 19: 55-63

Subject Index

MIX
Papier aus verantwortungsvollen Quellen
Paper from responsible sources
FSC® C105338

If you have any concerns about our products,
you can contact us on
ProductSafety@springernature.com

In case Publisher is established outside the EU,
the EU authorized representative is:
Springer Nature Customer Service Center GmbH
Europaplatz 3, 69115 Heidelberg, Germany

Printed by Libri Plureos GmbH
in Hamburg, Germany